ESTIMATING WATER USE
IN THE UNITED STATES

A New Paradigm for the National Water-Use Information Program

Committee on USGS Water Resources Research
Water Science and Technology Board
Division on Earth and Life Studies
National Research Council

D1491376

National Academy Press
Washington, D.C.

NATIONAL ACADEMY PRESS • **2101 Constitution Avenue, N.W.** • **Washington, DC 20418**

NOTICE: The project that is the subject of this report was approved by the Governing Board of the National Research Council, whose members are drawn from the councils of the National Academy of Sciences, the National Academy of Engineering, and the Institute of Medicine. The members of the committee responsible for the report were chosen for their special competences and with regard for appropriate balance.

Support for this project was provided by the U.S. Department of Interior and U.S. Geological Survey under Cooperative Agreement No. 01HQAG0030. The views and conclusions contained in this document are those of the authors and should not be interpreted as necessarily representing the official policies, either expressed or implied, of the U.S. government.

International Standard Book Number: 0-309-08483-0

Estimating Water Use in the United States: A New Paradigm for the National Water-Use Information Program is available from the National Academy Press, 2101 Constitution Avenue, N.W., Washington, DC 20418, (800) 624-6242 or (202) 334-3313 (in the Washington metropolitan area); Internet http://www.nap.edu.

Cover: Conceptual landscape design provided by John M. Evans, U.S. Geological Survey.

Printed in the United States of America

THE NATIONAL ACADEMIES

National Academy of Sciences
National Academy of Engineering
Institute of Medicine
National Research Council

The National Academy of Sciences is a private, nonprofit, self-perpetuating society of distinguished scholars engaged in scientific and engineering research, dedicated to the furtherance of science and technology and to their use for the general welfare. Upon the authority of the charter granted to it by the Congress in 1863, the Academy has a mandate that requires it to advise the federal government on scientific and technical matters. Dr. Bruce M. Alberts is president of the National Academy of Sciences.

The **National Academy of Engineering** was established in 1964, under the charter of the National Academy of Sciences, as a parallel organization of outstanding engineers. It is autonomous in its administration and in the selection of its members, sharing with the National Academy of Sciences the responsibility for advising the federal government. The National Academy of Engineering also sponsors engineering programs aimed at meeting national needs, encourages education and research, and recognizes the superior achievement of engineers. Dr. Wm. A. Wulf is president of the National Academy of Engineering.

The **Institute of Medicine** was established in 1970 by the National Academy of Sciences to secure the services of eminent members of appropriate professions in the examination of policy matters pertaining to the health of the public. The Institute acts under the responsibility given to the National Academy of Sciences by its congressional charter to be an adviser to the federal government and, upon its own initiative, to identify issues of medical care, research, and education. Dr. Harvey V. Fineberg is president of the Institute of Medicine.

The **National Research Council** was organized by the National Academy of Sciences in 1916 to associate the broad community of science and technology with the Academy's purposes of furthering knowledge and advising the federal government. Functioning in accordance with general policies determined by the Academy, the Council has become the principal operating agency of both the National Academy of Sciences and the National Academy of Engineering in providing services to the government, the public, and the scientific and engineering communities. The Council is administered jointly by both Academies and the Institute of Medicine. Dr. Bruce M. Alberts and Dr. Wm. A. Wulf are chairman and vice chairman, respectively, of the National Research Council.

COMMITTEE ON U.S. GEOLOGICAL SURVEY
WATER RESOURCES RESEARCH

DAVID R. MAIDMENT, *Chair,* The University of Texas, Austin
KENNETH R. BRADBURY, *Chair*, Wisconsin Geological and Natural
    History Survey, Madison (through December 2000)
A. ALLEN BRADLEY, University of Iowa
VICTOR R. BAKER, University of Arizona, Tucson (through December 2000)
ANA P. BARROS, Harvard University, Cambridge, Massachusetts (through
    December 2000)
MICHAEL E. CAMPANA, University of New Mexico, Albuquerque (through
    December 2001)
BENEDYKT DZIEGIELEWSKI, Southern Illinois University at Carbondale
N. LEROY POFF, Colorado State University
KAREN L. PRESTEGAARD, University of Maryland, College Park
STUART S. SCHWARTZ, University of North Carolina
DONALD I. SIEGEL, Syracuse University, Syracuse, New York
VERNON L. SNOEYINK, University of Illinois at Urbana-Champaign
    (through December 2001)
MARY W. STOERTZ, Ohio University, Athens
KAY D. THOMPSON, Washington University, St. Louis, Missouri

Staff

WILLIAM S. LOGAN, Project Director
ANITA A. HALL, Project Assistant

Editor

RHONDA BITTERLI

Preface

Water use is the aspect of water science most intimately associated with human activity. Study of water use is vital to understanding human impact on water and ecological resources and to assessing whether available surface and groundwater supplies will be adequate to meet future needs. To Native Americans of the arid West, water is not a commodity to be used, not the source of life—water is life itself.

Across the United States, the practices for collecting water use data vary significantly from state to state and vary also from one water use category to another, in response to the laws regulating water use and interest in water use data as an input for water management. However, many rich bodies of water use data exist at the state level, and an outstanding opportunity exists for assembling and statistically analyzing these data at the national level. This would lead to better techniques for water use estimation and to a greater capacity to link water use with its impact on water resources.

This report is a product of the Committee on Water Resources Research, which provides consensus advice to the Water Resources Division (WRD) of the USGS on scientific, research, and programmatic issues. The committee works under the auspices of the Water Science and Technology Board of the National Research Council (NRC). The committee considers a variety of topics that are important scientifically and programmatically to the USGS and the nation and issues reports when appropriate.

This report concerns the National Water-Use Information Program (NWUIP). The first national water use report published by the USGS appeared in 1950. The NWUIP itself began in 1978. The program operates continuously at the state or district office level, with a national summary integrating water use data from all the states produced every five years.

The work of the USGS in this area is important to many individuals and agencies working in water-related and other fields. Users of water use data and information range from local consultants and municipalities to natural resource economists, and from academic institutions to federal agencies such as the Department of Agriculture and members of Congress. Groundwater and surface water hydrologists from the USGS also depend heavily on this information to complete their water balances of aquifers and river basins.

Society's ever-growing need to understand and effectively manage its water resources has recently converged with the development and popularization of new tools for mapping and data analysis—tools such as geographic information systems and global positioning systems. These developments make this an exciting time to consider the possibilities for new directions for a program that has served the nation well in the past, but needs some reorientation to continue to serve the nation.

The committee heard the first presentations on this topic in October 1999. During the next 24 months, the committee met with numerous water-use experts from within and outside the USGS. The list of people who have helped us to formulate our ideas is long. Within the USGS, we are particularly grateful for the cooperation of the four regional water use specialists. Susan Hutson, Joan Kenny, Deborah Lumia, and Molly Maupin organized meetings, gave presentations, worked on collecting the data on state programs that appear in Appendix A and elsewhere, and answered numerous questions with patience and enthusiasm. Other USGS personnel who generously contributed their time and ideas include Bill Alley, Walt Aucott, Todd Augenstein, Nancy Barber, Jim Crompton, Scott Gain, Heidi Hadley, Bob Hirsch, Terry Holland, Marilee Horn, Pat Lambert, Gail Mallard, David Menzie, Bob Pierce, Eric Rodenburg, Greg Schwarz, Wayne Solley, Bill Templin, and Judy Wheeler.

We also appreciate the contributions of the many individuals from local, state, and federal agencies, academia, and elsewhere who gave presentations and participated in discussion sessions. These individuals include Ron Abramovich, Rick Allen, William Barron, John Boland, Emery Cleaves, Betsy Cody, Zena Cook, David Draughon, Ken Frederick, Noel Gollehon, Tom Huntzinger, Lane Letourneau, Earl Lewis, and Joe Spinazola.

Committee members then drafted individual contributions and deliberated as a group to achieve consensus on the content of this report. This report has been reviewed in draft form by individuals chosen for their diverse perspectives and technical expertise, in accordance with procedures approved by the NRC's Report Review Committee. The purpose of this independent review is to provide candid and critical comments that will assist the NRC in making its published report as sound as possible and that will ensure the report meets institutional standards for objectivity, evidence, and responsiveness to the study charge. The review comments and draft manuscript remain confidential to protect the integrity of the deliberative process. We wish to thank the following individuals for their review of this report:

John Boland, Johns Hopkins University
Peter Gleick, Pacific Institute
Noel Gollehon, U.S. Department of Agriculture
Thomas Huntzinger, Kansas Department of Agriculture
Marvin Jensen, Colorado State University (retired)
William Jury, University of California, Riverside
Scott Matyac, California Department of Water Resources
Joan Rose, University of South Florida

Although the reviewers listed above have provided many constructive comments and suggestions, they were not asked to endorse the conclusions or recommendations, nor did they see the final draft of the report before its release. The review of this report was overseen by Sandra Postel, Director, Global Water Policy Project. Appointed by the National Research Council, Ms. Postel was responsible for making certain that an independent examination of this report was carried out in accordance with institutional procedures and that all review comments were carefully considered. Responsibility for the final content of this report rests entirely with the authoring committee and the institution.

The committee wishes also to acknowledge the fine support and assistance that have been provided by Will Logan and Anita Hall of the Water Science and Technology Board staff. Will and Anita have organized our meetings very well, and done a great deal of work to transform a report manuscript into a final polished report. We want to acknowledge their contributions to the technical content of the report, but more than this, to express our appreciation on a personal level for their many contributions to the functioning of our committee.

We do not intend for this report to be seen as the "last word" in water use. Rather, it is hoped that the ideas generated in this report will stimulate further discussions, which need to take place not only within the USGS, but also with congressional staff, state and federal agencies, and other generators and users of water use data and information. We trust that these discussions will lead to new and better ways to integrate water use science into the human and natural world.

> David R. Maidment, Chair
> Committee on USGS Water Resources Research

Contents

Executive Summary

Water is fundamental to human existence. The United States has an extensive infrastructure for withdrawing surface and groundwater for public water supply, industrial and commercial use, irrigated agriculture, and livestock and domestic use and for cooling thermal power plants. The objectives of the U.S. Geological Survey (USGS) National Water-Use Information Program (NWUIP) are to quantify the nation's use of water and to develop and disseminate water use information at the local and national levels. To meet these objectives, USGS water use specialists work with state and local agencies that collect water use data for many purposes, such as supporting regulatory programs, fostering better water management, and determining of customer charges for water.

The term *water use* refers to all instream and offstream uses of water for human purposes from any water source. *Instream* use is a water use that takes place without water being withdrawn from surface or groundwater. *Offstream* use is use of water that is diverted from surface water sources or withdrawn from groundwater sources (a *withdrawal* in either case) and is conveyed to the place of use. This water is either lost to the system (*consumptive* use) or returned to surface or groundwater bodies (*return flow*), possibly with losses in transit (*conveyance loss*). Between withdrawal and return, the water may be *delivered* (to a public supplier, a water user, or a wastewater treatment plant) and then *released* one or more times.

NWUIP's principal product is a national summary of water use, produced every five years, that synthesizes county-level data for a five-year period. The year 2000 report will be the 11th five-year summary of this 50-year series. These reports provide the only nationally consistent, policy-relevant information on the status and trends of water use for the country.

1

However, the NWUIP faces many challenges. The NWUIP is the only USGS water resources program in which the Survey does not have the principal responsibility for primary data collection. Instead, the USGS relies on data of mixed quality, collected by many different organizations. For many of these data sources, the Survey can neither modify nor control the quality and accuracy of these data. The NWUIP is largely funded through the USGS Cooperative Water (Coop) Program. These funds are generally unavailable in states making little effort to collect water use data. It is therefore not surprising that the quality of water use data varies considerably from state to state.

Recognizing these challenges as well as the growing need for consistent national water use information, the USGS asked the National Research Council (NRC) to help determine the best approach for the future of the NWUIP. This report responds to that request.

In carrying out this study, the NRC Committee on USGS Water Resources Research considered basic questions about the nature of water use, the water use information needed in the United States, and the USGS's role in generating and disseminating that information. By studying state water use data collection and estimation programs, by analyzing data from the national water use summaries, and by examining parallel activities of other federal agencies, the committee arrived at a new vision for the future NWUIP.

GOALS FOR WATER USE INFORMATION

What kind of national water use information program is needed, and what role should the USGS play in that program? The committee concluded that such a program is needed in the United States to meet three primary goals:

1. *Maintain a comprehensive national water inventory.* With neither management nor regulatory responsibilities, the USGS is the only federal source of unbiased, science-based water resource information. The USGS inventories the nation's surface and groundwater resources and studies the processes affecting them. A national water use information program is essential for understand the effects of spatial and temporal patterns of water use on the quality, availability, and sustainable use of these resources.

2. *Help assure the nation's water supply.* The USGS's contribution to maintaining the nation's water supply requires continued focus on water use. Some of the most rapidly growing states, such as Nevada, are the most arid. Periodic droughts strain the capacity of water-supply systems, and climate change increases the uncertainty of future water conditions. Science-based assessments of the nation's water supply are therefore essential, and this need will become more intense for management and policy decisions in the future.

3. *Help preserve water quality and protect ecological resources.* The quality and quantity of water are inseparable; both profoundly influence aquatic

ecosystems. An example is wastewater discharge, which alters the receiving water quality and affects the downstream environment. The USGS's role in protecting aquatic and riparian ecosystems links its Water Resources and Biological Resources Divisions and requires detailed knowledge of water use.

The NWUIP is the nation's only comprehensive source of information on the status and trends of water use. The USGS is uniquely suited to provide this essential national information, working with its state-level cooperative partners and with other federal agencies. The committee identified substantial opportunities to enhance the activities and relevance of the NWUIP through a conceptual framework that emphasizes the integration of water use with the USGS's water resource programs.

CONCEPTUAL FRAMEWORK

It is essential to define an appropriate conceptual framework that links water use to its impact on water resources. The principal product of the NWUIP has been its five-year national estimates of *aggregated* water use, compiled primarily at the county and state levels. However, underlying these aggregated estimates is a more detailed *site-specific* description of water use, wherein each water withdrawal location may be tabulated, characterized, and quantified. At this level, the source of surface water or groundwater at each water use site can be identified, establishing the link between water use and water resources. The committee was impressed by the increasing availability of geospatial information on water resources generated by state regulatory programs, the USGS, and other federal agencies. In these programs, tabular data inventories describing water facilities are converted into mapped locations with attached attributes. This trend represents a rich and timely opportunity for systematically incorporating within the NWUIP consistent state and national data on water-using facilities.

These considerations led the committee to a conceptual framework for the NWUIP depicted in Figure ES.1. The physical system within which water use takes place is termed herein the *infrastructure water system*, described by locations of water withdrawals, by water discharges, and by the water facilities (pumping stations, treatment plants, water conveyance systems) by which water moves through the landscape in constructed water systems. The *natural water system* of streams, rivers, lakes, aquifers, and watersheds coexists with the infrastructure water system, and exchanges of water occur between them, primarily where water is withdrawn and discharged. In this framework, the NWUIP would be supported by water use data, water use estimation, and integrative water use science. *Water use data* are measurements or estimates of the amount of water used at a site or for a region. *Water use estimation* consists of random sampling methods, statistical inference, and other indirect methods to estimate water use at a site or for a region. *Integrative water use science* refers to the hypothesis-driven investigation of the behaviors and phenomena that determine spatial and

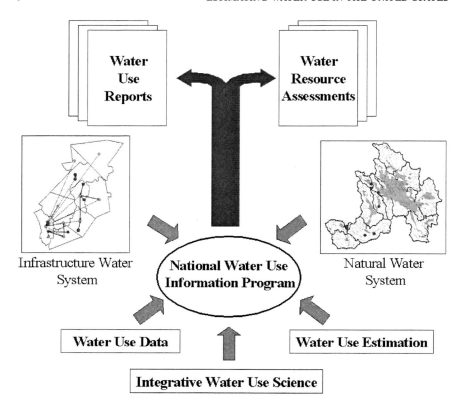

FIGURE ES.1 Conceptual framework for the National Water-Use Information Program. Note: The vignettes of the infrastructure and natural water systems shown here are drawn from a region in central New Hampshire using data from the USGS New England Water Use Database System.

temporal patterns of water use. It also includes scientifically assessing the impacts of water use on aquatic ecosystems, on the hydrologic cycle, and on the sustainability and vulnerability of the nation's water resources.

Within the committee's framework, the NWUIP would have two broad types of products: water use reports and water resource assessments. *Water use reports* (such as the five-year national summaries) include descriptive data compilations and summaries and status and trends information at the state and regional levels. *Water resource assessments* characterize the impacts of water use on the reliability and sustainability of groundwater and surface water resources and their associated aquatic and riparian ecosystems.

A GREATER EMPHASIS ON SCIENCE

The NWUIP appears to be viewed as a water use *accounting* program by many in the USGS Water Resources Division. Because of the significant differences in water use data collection procedures and in data quality from state to state, water use accounting alone cannot provide the estimates of water use that are needed by the nation. The committee found a compelling national need for unbiased science-based water use information. Long-term planning and management decisions need water use information that shows that water is an essential economic commodity and a vital natural resource. Investigations of the reliability and sustainable use of the nation's water resources require a fundamental understanding of the role of water use in the hydrologic cycle. The NWUIP should therefore be viewed as much more than a data collection and database management program.

The scientific understanding of water use will be most effectively achieved by making water use science the focus of the NWUIP. *Water use science* refers to the hypothesis-driven investigation of the behavior and phenomena that determine spatial and temporal patterns of water use. This science will directly contribute to the development of techniques that improve water use estimation. Water use science also includes scientific assessment of the sustainability of water resources and the impact of water use on aquatic ecosystems, on the hydrologic cycle, and on the reliability and vulnerability of the nation's water resources.

Recommendation: The NWUIP should be elevated to *a water use science* program, emphasizing applied research and techniques development in the statistical estimation of water use and the determinants and impacts of water-using behaviors.

WATER USE DATA

Water use data are collected throughout the nation to support the operation of water supply utilities and water districts. The types of data collected for other water uses vary significantly from state to state.

More than 20 states maintain comprehensive site-specific water use databases, most commonly developed to support regulatory programs that register or permit water withdrawals. In many cases these data are developed through cooperative projects between state water agencies and the USGS. In the remaining states, data are collected only for a subset of water use categories or areas within the states. Some states have no state-level programs for water use data collection. Water use data are more widely collected for surface water than for groundwater, and in some states (e.g., Arizona, Texas, Idaho, and New York), groundwater use data collection is focused on critical groundwater management areas.

There are several reliable national datasets with consistent information that can be used to identify and characterize water withdrawals. At the state level, information exists for more extensive regulatory functions than those conducted nationally, such as permit records for well drilling, water use appropriation, and the management and adjudication of water rights. Thus, even though water use data vary greatly from state to state, a great deal of information is available on water withdrawals in all states. The systematic synthesis and integration of these datasets is a key opportunity for the NWUIP.

The coupling of geographic information system (GIS) technology with these relevant databases is another major new opportunity for the NWUIP. Now, the NWUIP's only national compilation of water use information is aggregated water use data for counties, states, hydrologic basins, or aquifers. A national synthesis and integration of water use datasets is increasingly feasible, including data on the location and characteristics of water withdrawal sites and, where available, the amount of water withdrawn at the site. Statistical methods for rigorous sampling and estimation of water use for counties, states, hydrologic basins, and aquifers could be devised from such datasets. State and national datasets for water withdrawal by power plants, wells, and water treatment facilities are increasingly available. Other national datasets, such as county-level economic and employment data routinely compiled by the Bureau of Economic Analysis by SIC code, have consistent, national, aggregate information relevant to estimating water use. The synthesis of these datasets is increasingly feasible and is a timely opportunity for the NWUIP.

Recommendation: To better support water use science, the USGS should build on existing data collection efforts to systematically integrate datasets, including those maintained by other federal and state agencies, into datasets already maintained by the NWUIP.

WATER USE ESTIMATION

In practice, estimates of total water use for counties, states, watersheds, and aquifers are based on a combination of available data and surveys, supplemented by indirect estimation methods where survey data are unavailable. Consequently, the accuracy of the data summarized in the national water use reports varies both by state and water use category. The accuracy and confidence limits of these water use estimates are presently not quantified. Statistical methods can support more rigorous water use estimation. Such methods were explored by the committee in two illustrative examples for Arkansas data—stratified random sampling (SRS) and multiple regression analysis.

Using a high-quality, site-specific dataset containing every permitted withdrawal for the state of Arkansas, SRS with a well-constructed rigorous sample design was used to estimate total permitted withdrawals in the state. This example illustrates how SRS can optimally allocate limited sampling resources among homogeneous subpopulations, or "strata," of a well-defined statistical population. The number of samples selected from a given stratum is determined by the mean and variance of water use, the number of water use sites in the stratum, and the desired estimation accuracy. In the example presented in this report, water use categories (e.g., public water supply, irrigated agriculture) are used as strata. With additional population information, sample designs could incorporate other attributes such as geographic descriptors, climatic characteristics, or political subpopulations such as counties.

The committee's investigation of this Arkansas dataset showed SRS can produce reasonably accurate direct estimates of permitted withdrawals, and these results suggest that a rigorous comprehensive evaluation of the use of statistical sampling and estimation methods in the NWUIP may have value. Other states that similarly maintain comprehensive water use databases provide natural experimental settings for the systematic evaluation of these methods. As with all stratified random sample designs, the reduced sampling effort must be accompanied by the additional effort required to maintain current, accurate information about the underlying statistical population being sampled.

The committee also analyzed the structure of the 1980–1995 state-level data from the NWUIP by multiple regression analysis in order to determine if aggregate water use could be correlated with routinely collected demographic, economic, and climatic data. Indeed, a large number of potential explanatory variables for water withdrawal were identified. Potential explanatory variables include water price and gross state product for public water supply withdrawals and the existence of "closed-loop" systems (i.e., cooling towers) for thermoelectric withdrawals. Significantly, the analysis also found discernable responses to climate anomalies (such as droughts and heat waves) in aggregate state-level data; this suggests it is feasible to develop statistical techniques to normalize water use estimates for climate anomalies.

This analysis suggests how statistical methods may yield rigorous estimates and quantitative confidence limits for aggregate water use. Where cost-effective, water use estimates should be based on a complete inventory of significant water use sites in a region. In practice, water use estimates compiled for each water use category and for each county in the nation will consist of a combination of direct observation, random sampling, modeling, and statistical estimation. This mixture of water use estimation techniques requires a framework that best combines data from direct inventories with estimates from statistical sampling and modeling.

Recommendation: The USGS should systematically compare water use estimation methods to identify the techniques best suited to the requirements and limitations of the NWUIP. One goal of this comparison should be to determine the standard error for every water use estimate.

INTEGRATIVE WATER USE SCIENCE

The growth and development of water use science as a cornerstone of the NWUIP inevitably leads to a scientific synthesis, integrating water use with water resources. Integration of several different kinds can be done, including the following:

- integrating water use with water flow and water quality to develop a total picture of water moving through the landscape,
- integrating ecological uses of water within streams and aquifers as a component of water use, and
- integrating water use with those factors that influence it, including economic behavior.

The integration of water use and water resources envisioned here will require more data than just the location and quantity of water withdrawals. Integrative water use science will require a more complete data paradigm, capable of tracking the sources and fate of water use, including consumptive use, as it flows between the infrastructure and natural water systems. The New England Water-Use Database System, used by the USGS in New Hampshire and Vermont, is one useful approach. This is "a link-node" representation of the movement of water from points of withdrawal, through the infrastructure water system, to points of water discharge back to the natural water system. This approach to water use data collection, which is more comprehensive and integrated than that used in most other states, creates the data structures and framework needed to track changes in both water quantity and quality as water moves through the landscape.

The intimate linkage between water quality and water quantity creates both opportunity and need for better connections and integration among the NWUIP, the National Water-Quality Assessment (NAWQA) Program, and the Biological Resources Division (BRD) of the USGS. Water requirements for ecological uses of instream water compete with water withdrawals from streams for other uses. In integrative water use science, both withdrawal uses and instream uses are logically parts of the total water use, within which the effects of human water uses on ecological resources can be studied. The linkage of the infrastructure of water use with the natural water system offers a robust framework from which to understand the impacts of water use on water quality, aquatic ecosystems, the hydrologic cycle, and the reliability and vulnerability of the nation's water resources.

Recommendation: The USGS should focus on the scientific integration of water use, water flow, and water quality in order to expand knowledge and generate policy-relevant information about human impacts on both water and ecological resources.

FUNDING OF THE NWUIP

A significant limitation of the NWUIP has been its dependence on funding from the Coop Program. The quality and consistency of the NWUIP is uneven, reflecting the interests and cost-shared contributions of each state. Water flows through river basins and aquifers without regard to state boundaries. The national need to assure water supply and to protect water quality and ecological resources requires a stronger *national* focus on water use within the USGS. Implementing the major recommendations of this report will require continuity and dedicated support for NWUIP at the national level, analogous to the nationally funded component of the National Streamflow Information Program and of other USGS programs.

This report envisions a substantial transformation of the NWUIP. The committee does not see the NWUIP as simply a data collection and database management program focused on county-level categorical water use. Rather, the committee finds that a natural role for the NWUIP would be to complement and become actively integrated with the USGS's other efforts to provide unbiased science-based information about the adequacy and sustainability of the nation's water resources. Discussions within the USGS have identified the need to seek dedicated funds to support a national component of the NWUIP, and the committee endorses such a request. The synthesis and integration of national datasets will require active interagency liaison and ongoing dedicated staffing. A rigorous research effort focused on the intercomparison of statistical estimation methods is ideally suited to the USGS's research expertise and the nationwide infrastructure of the NWUIP. The active collaboration of hydrologists, hydrogeologists, biologists, and water-quality specialists is needed if water use is to be thoroughly integrated with the study of water flow, water quality, and ecological resources.

Recommendation: The USGS should seek support from Congress for dedicated funding of a national component of the recommended water use science program to supplement the existing funding in the Coop Program.

1

Introduction

The purpose of this report is to evaluate the National Water-Use Information Program (NWUIP) of the U.S. Geological Survey (USGS). The NWUIP is the main source of information about water use in the nation. This information is used by Congress, governmental agencies, and increasingly by research scientists to evaluate water used by the nation and alterations in water use that are related to demographic, economic, climatic, and other changes. The summary of national water use produced by the NWUIP every five years is the principal source from which spatial and temporal trends of water use in the United States are derived.

The evaluation process used by the committee included information gathering, program assessment, and the preparation of conclusions and recommendations. Information about the water use program was obtained from the NWUIP Web site (http://water.usgs.gov/watuse), from USGS and other publications, from other (unpublished) documents, from formal presentations by USGS staff, from analyses of NWUIP data, from discussions with water use specialists in various USGS districts, and from discussions with agency specialists in other USGS resource evaluation programs. The program evaluation was conducted in context with a statement of task that was jointly agreed upon by the USGS and National Research Council (NRC) staff. This statement of task is provided below.

STATEMENT OF TASK: REVIEW OF THE NATIONAL WATER-USE INFORMATION PROGRAM

"The study will provide guidance to the USGS on development of an improved National Water-Use Information Program. Beginning with an assessment of the existing USGS water-use information program, the study will consider how the program could be optimized within existing resources, considering analysis methods, data quality and presentation of results. The study will also consider if additional resources should become available, what would be the priorities for investing those resources to improve the program.

"Some specific questions to be addressed by the study include the following:

(1) Are there better ways to estimate water use (e.g. surveys or other statistical approaches)?

(2) How might the best approach vary by category of water use?

(3) What are the relative merits of the various water-use categories?

(4) What is the value of information at different spatial scales – state, basin, county, aquifers, and hydrologic unit?

(5) What is the value of information at different time scales and how can data be normalized for climatic variability?

(6) What should be the relationship of other agencies and institutions (universities; federal, State and local agencies; etc) to this program?

(7) What are the priorities for interpretive products from the program?

(8) What research could the USGS carry out that would lead to an improved water-use science."

CONCEPTUAL FRAMEWORK

Throughout its deliberations, the committee noticed that the USGS has not defined a conceptual framework for water use that places water use data and the NWUIP in context with the other water resources programs of the USGS. From a scientific perspective, this conceptual framework is an important link between water use and its impact on water resources. The principal product of the NWUIP has been its five-year national estimates of *aggregated* water use, compiled primarily by counties and states. Underlying these aggregated estimates, however, are more detailed *site-specific* data about water use, in which each individual water withdrawal locations are tabulated, the types of water use determined, and the amounts of water use estimated. At this site-specific level of water use, the source of surface water or groundwater for each water use site is identified, and the link between water use and water resources is established. The committee has been impressed by the increasing availability of national geospatial information on water resources, attributable to efforts of the USGS and other federal agencies to convert tabular data inventories describing water facilities into mapped locations with attached attributes. This activity makes possible "place-based" or "location-based" analysis in which georeferenced data from disparate sources are synthesized into an integrated database that can support more comprehensive study than can the separate data sources taken alone.

All these considerations have led to a conceptual framework for the NWUIP, as shown in Figure 1.1. The USGS is the primary agency that measures the

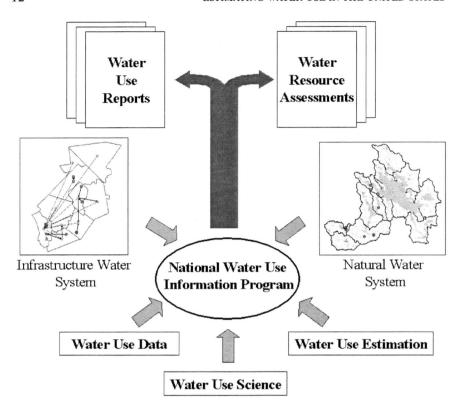

FIGURE 1.1 Conceptual framework for the National Water-Use Information Program.

quantity and quality of water in the *natural water system* of streams, rivers, lakes, aquifers, and watersheds. The natural water system coexists with the *infrastructure water system*, described by locations of water withdrawals and discharges, and by the principal water facilities (pumping stations, treatment plants, water conveyance systems) through which water is moved through the landscape in constructed water systems. Exchanges of water occur between natural water systems and infrastructure water systems, primarily at the points of water withdrawal and discharge. The principal focus of the NWUIP is the quantification of the *infrastructure water system*. The vignettes of the infrastructure and natural water systems shown in Figure 1.1 are drawn from a region in central New Hampshire using data from the New England Water-Use Database System, described in more detail in Chapter 7 of this report.

In this framework, the NWUIP is supported by water use data, water use estimation, and water use science. *Water use data* comprise the locations and characteristics of water withdrawals, discharges, and facilities and the amounts of water used at those locations. *Water use estimation* encompasses statistical sampling, regression, and other estimating methods such as coefficient models for determining the total water use within a given geographic region (often a site, city, or county) or from a particular water resource (e.g., a river or aquifer). *Integrative water use science* is founded on data and estimation, since these are the principal elements of observation of water use; however, it also includes integrating water use with water quality, and it includes environmental issues in water systems. The products of this conceptual framework are (1) *water use reports*, such as the five-year national water use summaries, and interpretive water use data reports prepared at the state and regional level and (2) *water resource assessments* in which the relationship between water use and availability of water is examined for particular aquifers or watersheds, or the impact of water use on water quality and environmental conditions is studied for particular regions.

DEFINITIONS OF "WATER USE" AND RELATED CONCEPTS

In this section, we briefly define terms that are commonly used in the USGS water use summary reports. The term *water use* refers to all instream and offstream uses of water for human purposes from any water source.

In the water use context, some terms acquire slightly different meanings than they have in the context water resources research. For example, in the water use context "instream" and "offstream" can both apply to either surface water or groundwater sources. The definitions below are from Solley et al. (1998), which includes an extensive water use glossary.

Offstream use refers to water that is diverted from surface water sources or withdrawn from groundwater sources (i.e., a "*withdrawal*") and is conveyed to the place of use. Most offstream or withdrawal uses that are compiled in USGS reports are self-supplied data from public suppliers, agencies, etc. Water use data are estimated for other categories, including domestic, commercial, irrigation, livestock, industrial, mining, and thermoelectric power.

Instream use refers to water use that takes place without water being withdrawn from surface water or groundwater. Instream uses include hydroelectric power generation, navigation, maintenance of minimum streamflows to support fish and wildlife, and minimum streamflows for wastewater assimilation. Hydroelectric power generation is the only instream use that is currently being evaluated on a national scale. California currently quantifies various types of instream uses.

Water that is diverted or withdrawn from surface or groundwater sources (offstream use) either is lost to the system (consumed) or returned to surface or groundwater bodies. These distinctions are made as follows:

Consumptive use refers to the portion of withdrawn water that is evaporated, transpired, or incorporated into products, animals, or crops. In some cases, consumptive use is evaluated as the difference between the volume of water delivered and the amount returned to water sources.

Conveyance loss is the amount of water that is lost in transit, either between the source and the point of use or from the point of use to the point of return. This is a category of consumptive use that has been separated out so that the effects of conveyance can be evaluated. Most of the conveyance loss is due to evaporation, along with seepage and leaks.

Return flow is the quantity of water that is returned to surface or groundwater sources. The quality of this water may be different from the initial water quality at the point of withdrawal.

Between withdrawal and return, the water may be *delivered* and then *released* several times. For example, after withdrawal from a stream, water may be delivered to and released from a potable water treatment plant, then a water user, and finally a wastewater treatment plant before being returned to the stream. The difference between these two amounts at any step is generally equal to the consumptive use.

OUTLINE OF THE REPORT

The report consists of nine chapters. This first chapter introduces the intellectual framework of the study and defines important concepts related to water use. Chapter 2 reviews the 50-year history of water use estimation by the USGS, the present state of the NWUIP, and procedures for water use data collection in each of the 50 states and Puerto Rico. Chapter 3 reviews water use data that are available as national datasets and on a state-by-state basis. Water use data from Arkansas are used in a case study to illustrate the characteristics of site-specific water use data. Water use data categories and metadata are also discussed in this chapter. Chapter 4 provides a review of the different tools and approaches that can be used to develop water use estimates using direct or indirect estimation techniques. Two such techniques are then discussed in greater detail. Chapter 5 looks at direct estimation through stratified (and unstratified) random sampling of the existing Arkansas dataset, for both groundwater and surface water. Chapter 6 examines indirect estimation using linear regression models that quantify the relationships between aggregated state-level water use data from the NWUIP and corresponding demographic, economic, and climatic data.

Chapter 7 reviews how the water use data and estimation procedures presented in previous chapters can be used to provide integrated water use information. The New England Water-Use Database System is used to show how water use and water resources data can be integrated in a particular region. Chapter 8 focuses on integrative water use science—i.e., the integration of water use data with related information on the quantity and quality of natural water systems.

This approach has the potential to answer questions about the consequences of water use on the natural water system and associated aquatic ecosystem, and it has the potential to evaluate whether the quantity and quality of water are sufficient for specific water use purposes. Finally, Chapter 9 presents the conclusions of the committee's study.

2

The National Water-Use Information Program:
Past and Present

Water use consists of withdrawal, conveyance, distribution, application, discharge, and reuse interwoven in a complex web of interrelated pathways and activities (Buchmiller et al., 2000). These uses are strongly influenced by environmental, economic, behavioral, and other natural and human society-based systems, as well as by the quality and quantity of the available water. Many consumers, including local and regional water management agencies, policymakers, scientists, teachers, and students, employ data on water use. The data are needed for water management planning because the national water supply is finite, and there is growing competition for this limited resource. Further, water management now involves more complexities than ever, including aquifer storage and recovery, artificial recharge, water reuse (both irrigation and wastewater), desalination (seawater and groundwater), and interbasin transfers. High-quality water use data are needed to establish water use trends and to forecast the effects of existing and contemplated policies. They are also needed to develop water use regulations to control undesirable trends, such as salt-water intrusion or groundwater overdraft, and to ensure maximum beneficial use of our water resources.

Water use data can also help us better understand the hydrologic and biogeochemical cycles and how humans impact these cycles. In particular, the influence of water use today on environmental systems is significant and will continue to increase (Buchmiller et al., 2000). Good water use data will assist policymakers in allocating funds for projects that ensure the proper balance between protection of the natural environment and societal use of the water resources.

In this chapter, we summarize the history of the National Water-Use Information Program (NWUIP), address the question of whether the U.S. Geological

Survey (USGS) should continue to be its coordinating agency, and outline some of the major challenges faced by the NWUIP. Finally, noting the dependence of the national program on data supplied by the individual states, we summarize the kinds of water use data collected by each of the 50 states.

HISTORY OF WATER-USE INFORMATION PROGRAMS
AT THE USGS

Information on water use in the United States before 1950 is limited. Water use data were collected and collated on an ad hoc basis by various federal agencies and other organizations with specific needs and objectives, such as the Natural Resources Conservation Service (NRCS), U.S. Bureau of the Census, U.S. Forest Service, U.S. Army Corps of Engineers (Corps), and U.S. Bureau of Reclamation (USBR). Picton (1952) estimated water use in the United States from 1900 to 1950, and Guyton (1950) estimated national groundwater use in 1945.

Estimated Use of Water in the United States – 1950 was the first report in what has become a 50-year series of five-year reports published by the USGS (MacKichan, 1951). This first report was followed by nine other reports from 1955 through 1995 (MacKichan, 1957; MacKichan and Kammerer, 1961; Murray, 1968; Murray and Reeves, 1972, 1977; Solley et al., 1983, 1988, 1993, and 1998) (Figure 2.1).

The 1950–1975 reports were compiled at USGS headquarters prior to the establishment of the NWUIP. Information was gathered from a variety of sources, including USGS district offices and state and federal agencies. It was recognized from the beginning that the water use data were subject to considerable error and were to be used only as estimates (MacKichan, 1951).

These USGS reports contained compilations and estimates of water use for various use categories for each state and water resources region. These data were further subdivided by surface water and groundwater use. The 1951 report created a template for the future reports; it made estimates for "withdrawal" uses, including municipal use, rural domestic and livestock use, irrigation use, industrial use from private sources, and water power. "Nonwithdrawal uses" such as navigation, waste disposal, recreation, and fish and wildlife also were qualitatively discussed (MacKichan, 1951).

Changes in the report format through 1975 were minimal. Consumptive use of water was estimated beginning in 1960, and some water uses were reclassified. However, the 1975 report was remarkably similar to the 1950 report, with categories including "(1) public supply (for domestic, commercial, and industrial uses), (2) rural (domestic and livestock), (3) irrigation, and (4) self-supplied industrial (including thermoelectric power)," along with hydroelectric power (Murray and Reeves, 1977). The potential use of the data for trend analysis of the different water use categories was recognized as early as the second report, which con-

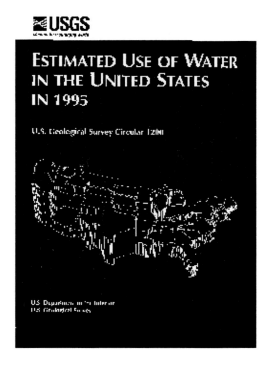

FIGURE 2.1   The USGS national water-use report for 1995.   SOURCE: Solley et al. (1998).

tained a table showing the percent change in withdrawals from 1950 to 1955 (MacKichan, 1957).   Figure 2.2 shows the trends in fresh groundwater and surface water withdrawals and in population for 1950–1995.

The NWUIP formally began in 1978.   Language approving the program appeared in the House version of the 1978 appropriations bill, which included "an increase . . . in the federal-state cooperative program [which includes] $1,000,000 to establish a national water use data activity."   The Senate concurred, and the conference committee made no further mention of the program.   The congressional language apparently did not specify the five-year time period that has become the norm for the USGS national water use compilations and reports (Wendy Norton, USGS, written communication, 2001).   The first full-time manager was soon appointed to the program, and by 1981, 47 states were participating in the program (Solley et al., 1983).

When the NWUIP was established, the USGS became concerned with data quality control and data management.   By 1980, the National Water-Use Data

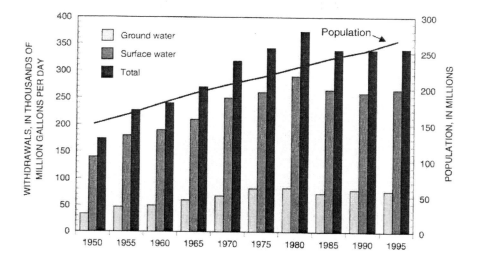

FIGURE 2.2 Trends in fresh groundwater and surface water withdrawals and population, 1950–1995. SOURCE: Solley et al. (1998).

System (NWUDS) was organized to store water withdrawal and use data by county and basin. This was later updated and renamed the Aggregated Water-Use Data System (AWUDS). By 1982, the Site-Specific Water-Use Data System (SWUDS) had been prepared and distributed to districts interested in using it. Detailed guidelines for water use report preparation were distributed (Lumia, 2000).

In the first decade or so of the NWUIP's existence, additional data elements and categories were added. For example, self-supplied industrial use was sub-divided into industrial, mining, and power (fossil fuel, geothermal, and nuclear) categories. Groundwater withdrawal reports specified the aquifer from which the water was extracted. Water use data were submitted by four-digit hydrologic unit or drainage basin. By 1990, estimates were made for thermoelectric power and wastewater releases, and consumptive use was estimated for all major water use categories (Solley et al., 1993).

However, financial and institutional pressures have forced the program to scale back the planned 2000 report (Figure 2.3). The water use categories for which data are being compiled at the county level for all states are public supply, industrial, thermoelectric, and irrigation. Mining, livestock, and aquaculture water use are obligatory only for the states where water use for these categories is large. Withdrawals from major aquifer systems are required only for public supply, irrigation, and industry. Commercial use, wastewater treatment, reservoir evaporation, and hydroelectric power are no longer being tracked, nor are

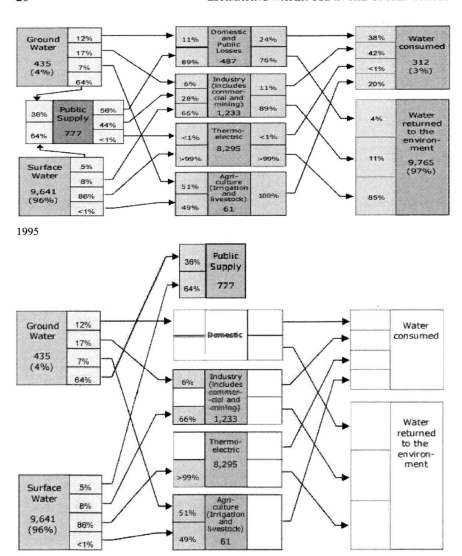

FIGURE 2.3 Comparison of data models (source, use, and disposition) for 1995 and 2000 national water use summaries. Data for unshaded boxes are not required for 2000 summary. Note loss of estimates for consumptive use and return flows. Volumes and percentages are for illustrative purposes only. Details on mandatory elements for different states are in Figure 3.2. Modified from NWUIP figures.

consumptive use, reclaimed wastewater, return flows, or deliveries from public suppliers.

The loss of many of the water use categories has obvious implications for local and regional studies of water availability and for studies of how human use of water impacts the quantity, quality, and sustainability of water resource systems, as described in Chapters 7 and 8. In brief, regional studies of water availability require the tracking of consumptive use and of locations and volumes of return flow such as wastewater discharge. Further, return flows must be linked to the stream or aquifer to which they discharge. Jurisdictions with complex water management and accounting concerns, such as aquifer storage and recovery, artificial recharge, water reuse, desalination, total maximum daily load (TMDL) issues, and/or interbasin transfers will not find the present categories sufficient for their needs. Many of these jurisdictions will continue to collect these data for their own purposes even though such data will no longer be tracked at the national level.

From the language of the 1978 House Appropriations bill, it is clear that the NWUIP was intended from the beginning to be within the USGS Cooperative Water (Coop) Program. This relationship provided a relatively constant source of funding, but it created structural limitations, which are discussed later in this chapter and elsewhere in the report. From fiscal year 1978 to 1985, federal funding for the water use program grew rapidly from $1.0 million to $5.0 million (USGS, 1981), and these funds were matched at 100 percent or more by the various state cooperators. However, from 1983 to the present, funding has remained fairly flat. The NWUIP is no longer specified as a line item in the budget submission to Congress (Lumia, 2000), but the federal share of the NWUIP funding is on the order of $4.5 million to $5 million (Wayne Solley, USGS, personal communication, 2000).

In the early years of the NWUIP, state cooperators were not required to provide their Coop Program match as direct funds, but were encouraged to enter the program through credits for "direct services" (i.e., credit for providing resources that can be used in the project, such as office space for a USGS water use specialist, supplies, etc.). Sometimes state cooperators were given "reverse flow credits" (Pierce, 1993). With this mechanism, a state matches the federal Coop contribution to pay for personnel to do the compilation. However, within a few years, the reverse flow credits were discontinued, and now even direct services matching is discouraged in most of the USGS regions (Lumia, 2000). This administrative structure has constrained the NWUIP in important ways, as is detailed later in this chapter.

SHOULD USGS CONTINUE TO ADMINISTER THE PROGRAM?

The NWUIP was established at the USGS, but it is important to consider whether this remains appropriate. Below we summarize some of the issues that must be taken into consideration.

The USGS is a nonregulatory agency. A number of federal agencies have strong mission-driven interests and technical expertise in water use. For example, the U.S. Army Corps of Engineers has a well-defined interest in water use supporting its national water resource planning and management responsibilities. Development of the IWR-MAIN water use model (Davis et al., 1991) was principally supported by the Corps. However, the Corps also regulates water use, including construction of water withdrawal structures and impoundments, in navigable waterways. These roles, which have traditionally emphasized municipal, industrial, and navigation uses, are principally regulatory, planning, and management—not scientific—roles.

Likewise, the U.S. Environmental Protection Agency (EPA) has strong regulatory interests in water use because it is the primary federal agency regulating drinking water and wastewater discharges. The EPA interest in water use is driven by risk management through wellhead protection and source-water protection programs and the Safe Drinking Water Act and by how water use information is used in risk assessments for exposure to toxic chemicals and waterborne pathogens.

Certainly these agencies can and do provide scientifically defensible data in many areas. But there have also been controversies, and the committee believes that these controversies are more likely where the data are reported by a regulatory agency. As water resources become increasingly stressed with an ever-increasing number of competing uses and users, it seems preferable to assure the public that water use estimates are as free from political influence as possible. Unlike the Corps, the EPA, and many other agencies, the USGS's interest in water use is wholly nonregulatory.

Water use relates to all major forms of economic activity. Various government agencies other than the USGS study water use for specific purposes. The most extensive of these efforts is that of the U.S. Department of Agriculture. The USDA's National Agricultural Statistics Service (NASS) conducts a Census of Agriculture (i.e., a "complete" inventory) every five years, including county-level information on number of acres irrigated by crop (USDA, 1999a). The Farm and Ranch Irrigation Survey (FRIS) (USDA, 1999b), a stratified random sample of about 10 percent of the Census farms, is then conducted one to two years after the Census and is reported at the state level. The FRIS collects data on water sources (groundwater, on-farm surface supply, and water from off-farm suppliers, which can include both ground and surface water), estimates of farm water applications (*not* withdrawals or consumptive use), and irrigation practices. The USDA Economic Research Service (ERS) uses data from the Census, the FRIS, and other agencies (e.g., USBR and USGS) for research on topics such as the economic value of water use and behavioral determinants of water use in agricultural production (Moore et al., 1994a, b; Schaible et al., 1995) and efficacy and equity of market mechanisms and policy actions (Anderson and Magleby, 1997; Morehart et al., 1999).

Although many aspects of the USDA program may serve as a model for the USGS water use program, the USDA program views water primarily as an economic input for agricultural production. However useful this perspective may be, it is not broad enough for a generalized national water use program. Domestic, industrial, thermoelectric, and other kinds of water use may be just as important to track and to compare with each other.

Water use is an important component of the hydrologic cycle. Although water use may be viewed in purely economic terms (see Chapter 4 for a discussion of input-output and materials flow analysis by the Bureau of Economic Analysis [BEA] and others), it may also be seen as the anthropogenic component of the hydrologic cycle. From this perspective, water use is as fundamental a part of the water cycle as precipitation, evaporation, and groundwater recharge and discharge. Users such as water-supply companies, consultants, and state and local government would naturally come to the USGS for their other hydrologic data and information, and it seems reasonable that the USGS would supply them with water use data as well.

The USGS has experience with hydrologic and hydrographic databases. The USGS administers the National Hydrography Dataset (NHD)—a set of digital spatial data on surface water features such as lakes, ponds, streams, rivers, springs, and wells. The NHD combines USGS national map information with river network information from the EPA. Surface water features are combined to form "river reaches," which provide the framework for linking water-related data to the NHD surface water drainage network. Water use data, including points of intake and National Pollutant Discharge Elimination System (NPDES) locations, can potentially be integrated with this dataset, as could water quality data from the USGS or the EPA. This integration would create a powerful tool for evaluating future water quality and quantity trends.

The USGS also operates NWIS (National Water Information System)—available on the Internet. The NWIS is a data system that includes real-time and historic data on streamflow and stage, groundwater levels, and water quality. Site-specific and aggregated water use data are also part of the NWIS system, although they are not yet well integrated with the other data types. The Site-Specific Water-Use Data System (SWUDS), will soon be fully compatible with other NWIS programs, and it should be ready for the NWIS 4.2 release in the spring of 2002 (T. Augenstein, USGS, personal communication, 2001).

The USGS operates other resource assessment programs. Opportunities for interaction of the NWUIP with other assessment programs in the USGS are possible, including the Minerals Information Team, which collects, analyzes, and disseminates information on the domestic and international supply of and demand for minerals and mineral materials, and the National Oil and Gas Project, which periodically assesses oil and natural gas resources and the potential of growing the reserve supply of the United States.

In summary, water use can be viewed through many different lenses (see Chapter 4), and there may be no single location for a national water use program that is "ideal" for every possible use of such information. Thus, a case could be made to place the NWUIP in any of the other agencies mentioned in this section (EPA, Corps, USDA, BEA) and possibly others (e.g., Bureau of the Census, USBR). Nevertheless, on the whole, the USGS's traditional role as the nation's nonregulatory, unbiased source of earth science data and information for multiple uses and users makes it a good choice to continue to run such a program.

However, as will be made clear later in this report, there is much that can be learned from other resource inventory programs within the federal government. The committee articulates a vision in which the water use program would remain within the USGS, but would maintain strong links to other programs such as the Farm and Ranch Irrigation Survey of the USDA. The water use program would have strong management and data links to USGS programs in groundwater resources, stream gaging, and water quality, but it would also maximize collaboration and coordination with other agencies that collect and store agricultural, industrial, demographic, and land use data, especially at the county level.

CHALLENGES FACED BY THE CURRENT PROGRAM

The USGS has achieved considerable success with the NWUIP and its predecessors over the past 50 years. It is easy to lose sight of the fact that half a century ago, the country knew very little about its use of such a vital resource. The institutionalization of the NWUIP in 1978, in particular, was a major step forward.

However, like all programs, it has its limitations and challenges. Before designing a vision for the future, it is important to understand these issues.

The Cooperative Water (Coop) Program funding mechanism. A major constraint on the NWUIP is that the program was created within the Coop Program and has remained there. The Coop Program depends on matching funds from a state or local cooperator. This brings inherent limitations. States that place a high premium on collecting water use information tend to develop a high-quality water use program, generally in collaboration with the NWUIP. Other states do not. Federal funds that could be used directly for water use programs are either not spent at all in that state or are used for programs of less direct application to water use, such as stream gaging.

The lack of a strong, stable headquarters staff. There have been, over the years, a number of dedicated NWUIP staff working for USGS headquarters. These people have historically coordinated the national compilation, organized water use training courses, developed and maintained site-specific and aggregated databases, performed limited quality control, communicated with Congress and with other USGS programs, and performed a wide range of other activities. However, the number of full-time staff at headquarters has only occasionally exceeded two persons and has sometimes consisted of a single individual. Recently, even that position has been eliminated, with additional responsibilities being placed on the four regional water use specialists and the chief of the Office of Ground Water. The limited number of national staff makes it difficult to do strategic planning, develop new estimation techniques, apply strategies used in one district to another nearby district, upgrade databases, and perform numerous other tasks necessary to long-term planning and management.

Disparities in data availability and quality. The disparity in spending from district to district has the immediate and obvious consequence that the quality and quantity of data available vary greatly from state to state. This committee firmly believes in the value of high-quality data; however, it is also aware that each state collects exactly as much data as it believes it needs and is willing to pay for. This creates problems at the aggregate level. The USGS, because of its broader national mission, has the responsibility to make reasonable estimates for water use even in states where little data exist and in states where there is no funding for a USGS water use specialist. The question arises as to who will pay the salary of the person who must make these estimates, at least once every five years for the national compilation. Whether this funding comes out of district-level "overhead" or elsewhere, there is an ad hoc nature to the process that probably does not serve the program well.

The consequences of this process are twofold. First, there may be a regional or national interest in state-level data. River basins and aquifer systems cross state lines, and watershed management may be complicated by poor information in major parts of a basin. Interstate water transfers are common, and the legal and regulatory framework may be dependent upon accurate data from all of the states involved. National issues such as predicting and mitigating the effects of global climate change require baseline data over broad regions.

Second, at the regional and national levels, trend analysis is also complicated by these disparities. Well-documented, quality-assured data from one district are mingled with poorly documented data from another district, the accuracy of which is highly questionable and poorly constrained. It is difficult, therefore, to draw conclusions concerning the confidence intervals surrounding national estimates and to determine whether changes in estimated water use are real or are random variations within the standard error of the data estimates.

Water use data that are poorly integrated with other kinds of USGS water data. As noted earlier in this chapter, the USGS has a national database for water data—the National Water Information System, or NWIS. At present, real-time and historic data on streamflow and stage, groundwater levels, and water quality can all be easily downloaded from the system, but water use data are not available from the same platform. (Data aggregated at the county and watershed levels from the 1990 and 1995 surveys are available at http://water.usgs.gov/watuse.) Historically, it has generally been more difficult for users to access water use data than other kinds of water data. This situation will be partly solved with integration of the site-specific data system in the NWIS 4.2 release in 2002.

However, difficulties with the integration of site-specific and aggregate data will presumably remain. This is partly because the existing AWUDS is anti-quated, and the time and capital investment required to update it to a state-of-the-art level will be considerable. However, some issues are related to the various needs and limitations of the cooperators. SWUDS is only being used by a small number of state cooperators. Of those states that maintain site-specific data, many use commercial, off-the-shelf databases from which the data for SWUDS can be extracted as mandated. These commercial databases may also be used by other divisions within the state government, and therefore they provide a consistent platform between the various state agencies. They may also be frequently upgraded as desired.

Other cooperators have developed their own software. One example of this is the Arkansas Soil and Water Conservation Commission, which found SWUDS slow and cumbersome and which wanted to be able to input the information in the field rather than having to send the raw data to the USGS. Further, the Commission also wanted to add non-water-related information, such as tax assessments, into the same database. Other state cooperators, such as the Georgia Department of Natural Resources, found that they could not use SWUDS because it asks for well-by-well data, whereas in Georgia, the data come in on a facility basis (i.e., with multiple wells lumped together). Various New England states wish to track consumptive use and recycled water, which the current version of SWUDS does not do. Thus, it will be difficult to construct a single, consistent database that most or all of the districts will use.

Also, there are some states that maintain databases intermittently or not at all. In all states, water data that are received for archiving may come from other antiquated or one-of-a-kind systems (Box 2.1).

Multiyear delays in publishing data. The nature of the water use data gathering process is very different from that of a stream gage or groundwater well equipped with telemetry. Unlike gages and wells that can transmit virtually real-time information, the collection of water use data may involve a survey, which may then be compiled and analyzed, transmitted to the USGS district office, aggregated in various ways, transmitted to the regional water use specialist or headquarters, compiled nationally, and finally published in the five-year national summary. This may be a lengthy process. Although most applications of water

---

BOX 2.1
Case Study of an Anachronistic System

At the water treatment plant in an Ohio town of 40,000, water pumpage is recorded at the plant by hand-transcribing numbers every eight hours from a computer screen. The computer software was custom-designed by a consultant in Ohio. The software is "a little bit primitive," according to the plant operator, and data glitches occur frequently. When the operator recognizes an obvious data flaw, she is forced to omit data from the report to the state (Ohio Department of Water Resources, Division of Drinking and Ground Waters), leaving a gap in the pumping record. She sees a need for real-time, well-specific data for normal operations and a need for access to regional data for contingencies, including contamination and drought. She feels that the data collection system is anachronistic, given the available technology.

---

use data do not demand real-time availability, there are exceptions to this general rule such as times of drought or contamination emergencies. Clearly, some of the process of making the data available to users can be expedited by the USGS; however, other aspects will always be out of its control, depending on the individual state or local governments. Network linkage to a central database would clearly expedite the data-delivery process, but this raises issues of quality control, data ownership, financial cost, and possibly security. Chapter 6 discusses the potential for generating estimates for any year by making adjustments for factors such as climate.

Locally, great strides in real-time data collection have been made at local and even state levels. The fact that the producers are also users creates an incentive to improve access and delivery of data. As a result, many facilities, for their own purposes, already collect data digitally and then as a separate step aggregate it for mandatory reporting.

Paucity of metadata. The regional water use specialists do their best to assure quality of data, and some quality assurance/quality control is provided within AWUDS software and through regional and national reviews of the data and the documentation. However, the quality of methodology documentation is highly variable from state to state (Molly Maupin, USGS, personal communication, 2000), and historically there have been difficulties in ascertaining exactly how water use estimates were made in certain districts. Additionally, cooperators may use inconsistent categories and definitions, and various data transformations may be made by the USGS water use specialist. Consequently, it is often difficult for the data user to determine the quality of the data collection effort and to assign confidence intervals to the published numbers. This is problematic, especially if the data are being analyzed for lawsuits, for environmentally sensitive applications, or for trend analysis. Data issues are discussed in more detail in Chapter 3.

A wide range of data users. Earlier in this chapter, we described data-quality problems associated with the varied *sources* of water use data. The *users* of the data generated by the NWUIP, and their corresponding data needs, may be even more varied. Figure 2.4 summarizes some of this complex flow of water use data and information. Note that some entities are both suppliers and users of data. For example, a public water-supply district for a major city clearly represents a source of data for the NWUIP. This same institution may use irrigation water use trends for their region for forecasting water availability in the next decade.

The three primary user groups are the following:

1. those who need and collect data for their own purposes, such as billing, budgeting, allocation, resource management, lobbying, and accounting (this group includes the agencies listed in Table 2.1),

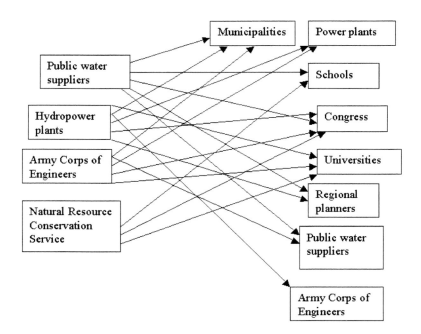

FIGURE 2.4 A representative subset of the suppliers and users of water use data. Note that some entities are both suppliers and users of data.

TABLE 2.1 Agencies that Need and Collect Water Use Data for Their Own Purposes

Public water treatment utilities	State cooperative extension services
Wastewater treatment utilities	Rural water associations
American Water Works Association	Agencies responsible for emergency planning
Power utilities	River basin commissions
Public Utilities Commissions	U.S. EPA (Publicly Owned Treatment Works)
Industrial facilities	U.S. EPA (NPDES compliance)
State agencies responsible for allocation or resource management	U.S. Army Corps of Engineers (hydroelectric)
State agencies responsible for compliance with Clean Water Act, Safe Drinking Water Act	U.S. Department of Energy, Energy Information Administration
State health departments	U.S. Bureau of Reclamation
State geological surveys	U.S. Bureau of Land Management
State agencies regulating mining	National Agricultural Statistics Service
State departments of agriculture	USDA's Natural Resources Conservation Service

2. decision and policy makers, including regional or state planning offices, legislative and executive branches of government, and state, federal, and local government (city councils, county commissioners, etc.), and

3. educational institutions, including elementary and high schools, colleges, and universities.

STATE WATER USE DATA COLLECTION PROGRAMS

The NWUIP would not be able to function as it is presently constituted if it were not for water use data collected by the individual state governments for their own purposes. This raises the question, "What kinds of data are collected and stored by each state, and how often?"

At the request of the committee, the USGS water use specialists undertook a survey of the current condition of water use data collection in all 50 states, the District of Columbia, and Puerto Rico to answer this question. The survey was significantly assisted by the four regional water use coordinators: Deborah Lumia (Northeastern Region), Joan Kenny (Central Region), Molly Maupin (Western Region), and Susan Hutson (Southeastern Region). The committee is indebted to these water use specialists and to their colleagues in each state for providing this information.

A series of questions were addressed to the water use specialist in each state, and the responses were tabulated into spreadsheet format by Ms. Lumia. The spreadsheet information was then summarized in narrative form for each state by the committee, and the resulting summary was checked for accuracy by the regional water use coordinators.

This survey should be considered only a first attempt to develop a comprehensive database of state water use policies. We also encourage follow-up studies on the impacts of these policies on local, regional, and national water resources and on water use patterns.

SURVEY QUESTIONS

The following nine questions were addressed to the USGS water use specialists in each state:

1. Does your state have the legal authority to permit or register withdrawals? (If NO for all categories, stop here).

2. Are withdrawals reported to the state? How often (e.g., annually, monthly) are withdrawals reported?

3. What is the trigger level above which the law applies? For example, 10,000 gallons/day, 100,000 gallons/day in any 30-day period, etc.

4. Is the law statewide or applicable only to certain parts of the state (counties or basins, for example)? Please indicate the applicable geographic area.

5. Do the laws differ for groundwater and surface-water withdrawals? If so, how?

6. Does your state maintain a database of reported withdrawals? If yes, how often is the database updated?

7. Does your state store the latitude and longitude of groundwater wells, surface water intakes, or other water use entities?

8. Does your state check the reported data against any other information? If so, what is used?

9. Does your state track changes to the reported data?

SURVEY RESULTS

A narrative describing the results of the survey in each state is presented in Appendix A. The general character of the responses to each question is summarized here.

1. There are two main purposes for collecting water use data: to support a water registration or permitting process and to support water resources management and planning. States may collect water use data for water management purposes even if they do not have the legal authority to permit water use. Public water supply is regulated by the EPA through state agencies using more consistent procedures than apply to other categories of water use.

2. The most complete site-specific water use data programs collect monthly water use and store it in a database that is updated annually. More typically, data on water use are collected and stored annually and are updated annually.

3. The most frequently cited trigger level for collecting water use data for all users is for average withdrawals exceeding 100,000 gallons per day. Other trigger levels used are 50,000 gallons per day, 10,000 gallons/day, and a variety of other discharge rates expressed in various units.

4. Most states treat water use uniformly throughout the state. However, a significant number of states have special data collection requirements for particular groundwater resource regions, such as the Eastern Snake Plain aquifer in Idaho, or Long Island, New York. Florida's five water management districts each have different reporting requirements, some based on well discharge and others on well diameter. In a few states (Colorado, Nevada, New Mexico), the state engineer has great latitude in issuing water permits and in deciding which permit holders must submit data on their withdrawals.

5. Laws governing water use are the same for surface water and groundwater in most states, but in some Western states, (e.g., Colorado, Nebraska, Texas), there is a significant distinction between water use data collection programs for surface water and those for groundwater. Data on surface water use are collected more universally than data on groundwater use.

6. Most states that collect data store the results in a database or have the USGS maintain their water use database. The database is usually updated annually, but in some cases, it is updated less frequently, such as every two or three years or when the permits are renewed.

7. Only a few states presently store the latitude and longitude of groundwater wells and surface water intakes. A larger number store the locations of wells and intakes as township-range-section values from the Public Land Survey System (PLSS; this is further discussed below). There is a trend toward storage of water use site locations using latitude-longitude coordinates checked by using global positioning system (GPS) equipment.

8. In many states, water use data are collected and stored but are not checked and verified against independent information. Where checks are done, the most frequent is to compare the current year's water use to the previous year's water use at each water use site. Some states check reported water use against the water use permit amount.

9. The most common tracking of water use data is to look at year-to-year differences.

WATER REGISTERS AND PERMITS

The first question on the survey—"Does your state have the legal authority to permit or register withdrawals?"—is important in defining the quality of the state water use data collection program. In a water use *register* system, a state does not issue permits for water use but does require users to report their water use amount, usually annually. In a water use *permit* system, a state requires permission to withdraw water from a stream or aquifer. Permits take two forms: (1) the site is permitted, but the total amount of the withdrawal is not specified, and (2) the site location and annual withdrawal amount are both specified in the water use permit. The significance of these distinctions was not apparent at the time the survey was undertaken, and in retrospect, it would have been better to clarify the type of register or permit system used by the state.

The results of this state water use data collection survey are illuminating, and they document in which states comprehensive site-specific water use data collection programs exist. However, the survey also revealed that each state has a particular water use environment, pattern of state laws governing water use, and history of water use data collection and interpretation by the USGS and state agencies. The information presented in this report is only a brief summary of a complex situation that deserves to be more fully documented. At several meetings where the committee met USGS and state water use specialists, many publications were presented to the committee describing water use trends and interpretations for particular states or regions.

TOWNSHIP-RANGE-SECTION COORDINATES

A number of states have described their water use sites with a location specified by township-range-section values. These values are based on a grid system called the Public Land Survey System (PLSS), a method of locating parcels of land. The method, originally developed in 1785, involves subdividing what were then the "public lands" of the United States— i.e., the land outside the original 13 colonial states (Minnick and Parrish, 1994). The PLSS applies now in all states except for the original 13 states, Texas, and Hawaii. A rectangular grid of "townships" is laid out over each state using one or more specified meridians and parallels as baselines. The township (east-west) and range (north-south) values are the coordinates of a Township cell in this grid. Townships are divided into 36 smaller grid units called sections, then into quarter sections, and finally into quarter-quarter sections. Thus, if a water use site is identified by township-range-section, it is located somewhere within a particular cell in the PLSS grid.

Though this system of labeling water use sites has the merit of providing at least some spatial location information, it is clear that labeling water use points by latitude and longitude is a preferable alternative that can be physically checked with a GPS unit. The water use site can then be more precisely located on a map than just within a grid cell in the PLSS. A particular shortcoming of the township-range-section system for a national water use database is that this system is not used in 15 states.

Finally, it should be noted that methods of varying degrees of accuracy have been developed by individual states and the private sector to convert PLSS descriptions to decimal degrees (i.e., latitude-longitude). However, there are numerous sources of error in such datasets and programs, and many water use site descriptions do not give northing and easting within the section or quarter section. Such sites would have location accuracies on the order of ± 2,640 feet (section) and ± 1,320 feet (quarter section), which is not sufficient for many purposes.

STATE WATER USE DATA COLLECTION CATEGORIES

The responses from this survey reveal a patchwork quilt of data collection, ranging from states such as Delaware, Hawaii, Kansas, and New Jersey, which have databases that are as complete as that in Arkansas, to states where little or no water use information is collected by state institutions. It seemed useful to the committee to try to classify the responses from the survey into some form in which the states could be categorized as to their degree of water use data collection. It is apparent from the survey that the strongest data collection programs are usually found in those states where the state has the legal authority to register or permit water withdrawals. However, for its five-year water use data summaries, the USGS collects data in all states, so there is a minimum level of data collection everywhere. In some instances, such as the Illinois State Water Survey, a strong water use data collection program exists even though the state does not register or permit water withdrawals. The state water use data collection programs are classified into the following three categories:

Category 1: The state has the legal authority to register or permit water withdrawals throughout the state for all water users from surface water or groundwater sources whose withdrawal rate exceeds a trigger level of 100,000 gallons per day. The state requires the collection of water use data, maintains a database of monthly water use updated annually for each withdrawal above its trigger level, and records water withdrawal locations using latitude and longitude coordinates. Some checking of data quality is carried out. These data are suitable for all water use reporting and for water availability studies.

States within Category 1 are Arkansas, Delaware, Hawaii, Indiana, Kansas, Louisiana, New Jersey, New Hampshire, Vermont and Massachusetts.

Category 2: Category 2 is the same as Category 1, but in Category 2, the data are recorded annually instead of monthly, and the withdrawal locations may be specified by township-range-section values rather than by latitude and longitude. These data are suitable for annual water use reporting but are less effective than date in Category 1 for water availability studies.

States within Category 2 are Alabama, Illinois, Maryland, Minnesota, Mississippi, New Mexico, North Dakota, Ohio, Oklahoma, Oregon, Utah, and Virginia.

Category 3: The state and/or the USGS collects and maintains water use data for some classes of users and/or some geographic areas within the state.

States within Category 3 are Alaska, Arizona, California, Colorado, Connecticut, District of Columbia, Florida, Georgia, Idaho, Iowa, Kentucky, Maine, Michigan, Missouri, Montana, Nebraska, Nevada, New York, North Carolina, Pennsylvania, Puerto Rico, Rhode Island, South Carolina, South Dakota, Tennessee, Texas, Washington, West Virginia, Wisconsin, and Wyoming.

A map of state water use data collection programs classified into these categories is presented in Figure 2.5. There are 10 collection programs in Category 1, with some tendency for these programs to be located in the central and eastern states. Thus, the Arkansas water use data collection program described in Chapter 3 is not an isolated outlier in terms of state water use data collection programs. The 22 states with data collection programs in either Category 1 or 2 have systematic programs of water use data collection in which both the usage amount and the coordinates of the location of usage are recorded in a database for all significant water use categories.

For the remaining 28 states, the District of Columbia, and Puerto Rico, water use data collection programs vary significantly. For example, Washington does not have a water use data collection program, Wyoming has a systematic program but not for all water use categories, and Pennsylvania has a systematic program but not for all areas of the state. It is worth noting that some of the most water-stressed states (e.g., California, Arizona, Nevada, and Colorado) fall into this category.

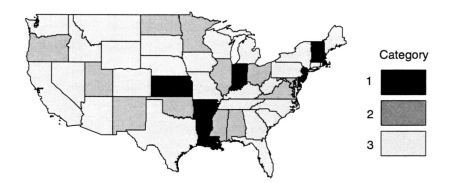

FIGURE 2.5. Classification of state water use data collection programs. See text for explanation of categories.

CONCLUSIONS AND RECOMMENDATIONS

Most of the committee's major conclusions and recommendations come later in the report. However, several conclusions and recommendations arise directly from the structure and limitations of the existing program:

• A national water use program is vital for effective water planning and management. High-quality water use information is critical to maintaining a comprehensive national water inventory, assuring the nation's water supply, assessing the effects of global change on water use, and preserving water quality and protecting ecological resources.

• The USGS is probably the most appropriate place to carry out such a program. Water use may be studied using many methods, including some methods (e.g., econometric and materials flow) in which the USGS Water Resources Division has limited experience. However, on the balance, the USGS's role as an unbiased provider of information, its expertise in resource assessment and hydrologic databases, and the important connection of water use data with the hydrologic cycle argue strongly in favor of keeping the program at the USGS.

• The overwhelming dependence of the NWUIP on the Coop Program funding (i.e., projects must attract 50 percent funding from local sources or they are not funded) is undesirable. *Like the stream gaging program, which has both Coop and national funding, the NWUIP requires a component of federal funding that does not depend on the interest level of individual states.*

• Whether done solely within the NWUIP or also through collaboration with other programs within the USGS, *a stronger focus on water use science at the federal and regional levels is imperative.*

• There are real disparities in the quality and quantity of water use data collected by the various states, and these disparities are likely to persist. Any national-scale program must find mechanisms to alert the user to these differences and to encourage states to collect the highest-quality data that are consistent with their needs and resources. Therefore, *as part of the current five-year national water use estimation effort, a separate publication should be prepared documenting the condition of the water use data collection program in each state and cataloging the publications that have been prepared to summarize and interpret these water use data.*

3

Water Use Data and Their Application

At its heart, a water use program is only as good as the quality of its data. This chapter is concerned with these data—the need for high-quality data, categories of data, and metadata and their value at various scales of space and time. Later in the chapter is a series of case studies based on a fairly comprehensive, site-specific set of data from Arkansas. These studies demonstrate how high-quality water use data can be used to show spatial and temporal patterns in groundwater and surface water use, water use intensity, and water stress. Finally, we end with a summary of some of the relevant federal databases that may be used to assist in making water use estimates, especially in states that have limited datasets of their own.

THE NEED FOR WATER USE DATA

Water use is part and parcel of the water budget. In much of the eastern half of the nation and in parts of the Northwest, where precipitation is greater than potential evapotranspiration, most water is used in the same basin from which it is withdrawn (with important exceptions such as New York City and Boston). In arid parts of the nation, precipitation is considerably less than potential evapotranspiration, and surface water and groundwater resources are often scarce. This has led to extensive mining of nonrenewable water in groundwater storage and to the importing of water from outside basin boundaries to meet water supply needs. Many regions rely heavily on water provided to reservoirs from melting winter mountain snow packs. These regional limitations on water availability have led to complex water rights laws and intrastate and interstate water rights legal battles. See, for example, Marc Reisner's book *Cadillac Desert: The American West and*

Its Disappearing Water (Reisner, 1986). Examples of western water problems are described in Boxes 3.1 and 3.2.

Much as debits, credits, and savings in a financial budget need to be quantified to maintain fiscal responsibility, the nation's water use needs to be comprehensively quantified within the water-budget context to ensure adequate availability of water as future water demands regionally fluctuate because of changes in climate, urban growth patterns, agricultural practices, and energy needs.

Many of the other components of the water budget are well studied. The National Oceanic and Atmospheric Administration (NOAA) monitors nationwide precipitation input—e.g., through the NEXRAD program (http://www. roc.noaa.gov/) and individual precipitation stations (http://www.websites. noaa.gov/guide/sciences/atmo/precip.html). Much of the nation's surface water flow is similarly monitored by the U.S. Geological Survey (USGS) at high temporal and significant spatial resolutions to allow projections to be made on flood frequency, discharge, river stage, and reservoir and lake storage. The USGS has also developed and continues to develop extensive regional datasets on changes in groundwater levels and the status of the nation's major aquifer systems (http:/ /water.usgs.gov/cgi/rasabiblio; http://water.usgs.gov/nwis). Comprehensive soils maps are available from the Natural Resources Conservation Service (NRCS), providing spatial data on important engineering properties such as infiltration rate and soil moisture storage.

The collection of water use data, however, has been given short shrift nationally with respect to funding and infrastructure support. Without consistent, comprehensive, and reliable water use data, appropriate decisions on water use management cannot be made. Well-structured water use data must be a part of the national water spatial data infrastructure if we hope to make sound future decisions regarding regional water allocation. (Further discussion of sources for such data is found in the section of this chapter titled "National Water Use Data: Federal Site-Specific Databases," and in Chapter 7.) The products of a national water use program will ultimately be multifaceted, helping to address how to maintain sustainable water in aquifers, sustainable water for industrial, agricultural and urban development, and sustainable baseflow in streams.

CATEGORIES OF WATER USE AND WITHDRAWALS

For its 2000 water use summary, the USGS has lumped water withdrawals into broad categories (Figure 3.2): public supply, domestic, industrial, thermoelectric, and irrigation. There is no longer an effort to track consumptive use, as was done for the 1995 water use summary. Also, there are no detailed breakdowns within specific categories, and return flows and instream water will not be documented as mandatory elements.

The rationale behind the lumping is based on the recognition that water use data effectively are collected by the USGS district program in an "ad hoc" man-

BOX 3.1
The Central Arizona Project

Combining low annual precipitation with extremely high evapotranspiration, the deserts of central and southern Arizona have always inspired a human concern for water. Although originally founded on the Santa Cruz River, the southern Arizona city of Tucson reduced that stream to ephemeral flood runoff because of groundwater pumpage in the 1940s. The solution to such overdraft had, since Arizona statehood, been envisioned as the long-distance transfer of Colorado River water. However, it was California that first achieved that dream, following dam construction in the 1930s. Arizona so resented California's access to the Colorado that it refused to join the six other Colorado Basin states in dividing up the river water through the 1922 Colorado River Compact. After a futile attempt by its National Guard to prevent California's extraction of Colorado River water, Arizona eventually joined the compact in 1944 and initiated a unified political effort to achieve its own engineered water-transfer system.

The U.S. Congress approved the Central Arizona Project in 1968, and in 1973, construction began on the 336-mile-long series of aqueducts, canals, pumping stations, and siphons. When federal funding to the project was threatened by the Carter Administration's criticism of the state's water conservation practices, the Arizona Legislature reluctantly passed the then innovative Groundwater Management Act of 1980. A state agency, the Arizona Department of Water Resources, was thereby created to manage and allocate both surface water and groundwater. Although the need for Colorado River water had been originally justified to supply agriculture and mines, the economy of the state changed radically over the decades. By the early 1990s, the project was ready to deliver surface water to Tucson, one of the largest U.S. cities solely dependent upon groundwater for its supply, but few farms or mines signed contracts for access.

Central Arizona Project water arrived in Tucson just in time to resolve a problem not envisioned at the time of project justification. Since World War II, the urban population had grown from less than 100,000 to over 800,000. The groundwater withdrawals were threatening future development by violating the conservation provisions enacted in 1980. By replacing groundwater with Colorado River water for municipal use, continued growth would be assured. Delivery of surface water began in November 1992. It was later recognized that a combination of old galvanized steel or iron water mains plus a different water chemistry from the original groundwater resulted in numerous cases of damage to water-using home appliances and also resulted in stained water and skin rashes. Public complaints about the water were poorly handled by the water delivery agency, and a political initiative eventually resulted in a November 1995 act that outlawed direct delivery of Central Arizona Project water to Tucson unless the water was treated to the quality of the original groundwater.

The irony of Arizona's long political struggle to bring Colorado River water to the desert, only to have it rejected by democratic vote, raises interesting questions about long-term planning, public understanding, and social factors in water supply. After a lengthy campaign of education, political advertising, and public relations, provisions of the 1995 act were revised in 1999 to allow the progressive municipal use by Tucson of water delivered by the Central Arizona Project. However, few expect this to be the final chapter in the continuing story of bringing water to the Arizona desert.

BOX 3.2
Use of Water Use Information in Interstate Arkansas River Water Litigation

Kansas and Colorado have fought over water for more than 100 years. In 1985, Kansas again sued Colorado, alleging that Colorado and its water users had depleted the usable and available flows of the Arkansas River (Figure 3.1) as measured at the Colorado-Kansas state line, in violation of the 1948 Arkansas River Compact. Among other claims, Kansas argued that flows had been depleted by about 1,500 post-Compact large-capacity wells drilled in Colorado, which pump groundwater in hydraulic connection with the Arkansas River.

Consideration of this case by the "Special Master," including consideration of the remedy and damage issues, involved the determination of Colorado's depletion to usable stateline flow for various time periods. Data for the Hydrologic Institutional Model used to simulate river flow included inputs to the Arkansas River from the western slope and all subsequent withdrawals and returns along the river to the state line. As part of the model, data were amassed on precipitation, on withdrawals for irrigation and other uses, on phreatophytic water use, on streamflows, and on return flows.

Streamflow information was obtained at USGS stream gaging stations, and information on metered surface water withdrawals for irrigation was available from irrigation ditch companies. However, irrigation well pumpage was more difficult to determine because few Colorado wells are metered. Information on crop acreage, crop water needs, and climatic factors was used to estimate irrigation groundwater use in Colorado. Electric power utility records were used to identify wells in use

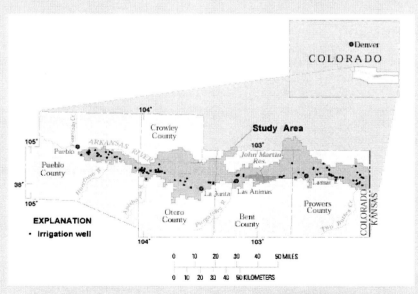

FIGURE 3.1 The lower Arkansas River in Colorado. Source: Dash et al. (1999).

BOX 3.2 Continued

and duration of their use, and coefficients were developed relating power consumption to amount of water pumped at metered wells.

In the end, the Special Master and, later, the U.S. Supreme Court concluded
that post-Compact well pumping in Colorado had violated the Compact. Kansas
will receive monetary damages for four costs: (1) additional pumping costs to
replace water Colorado did not deliver, (2) additional regional pumping costs associated with groundwater declines caused by this shortfall, (3) crop losses, and
(4) secondary economic effects. Data on water use and water levels in Kansas
have been used in conjunction with determinations of increased pumping costs to
Kansas farmers.

In summary, water use data, along with hydrologic, hydrogeologic, meteorological, economic, and power utility data, helped form the basis for a comprehensive
picture of the basin's water budget and for the resolution of a major interstate civil
lawsuit.

SOURCE: Joan Kenny, USGS, and Leland Rolff, Kansas Department of Agriculture.

---

ner and also on the recognition that the USGS necessarily has to rely largely on
water use data provided to it by state agencies and other parties. There is only a
limited national-level quality assurance/quality control program for most of the
received data. This makes it difficult to ascribe confidence intervals to the data,
although regional water-quality specialists in their reports have made some independent estimates of error in the various categories used in the past (Snavely,
1986). Fundamentally, water use must be treated with the same scientific rigor
used to quantify other aspects of the water budget, even if water use data inherently have much greater uncertainty. Because there now is large uncertainty in
many estimates of water use, it becomes even more important to develop the
scientific tools to address and minimize the uncertainty as much as possible.

As now designated, *public water* is water withdrawn by public water suppliers and delivered to users (USGS, 2000). These water suppliers (e.g., city well
fields) provide water mostly to domestic and commercial clients but also can
provide water to irrigation and to some industrial uses, including mining and
thermoelectric power generators. Public water supply data are commonly obtained
from providers. *Domestic water* is water used for drinking and household purposes. Usually, domestic water refers to water obtained from individual home
wells or surface water supply. However, because domestic water can also be
provided by public water supplies, it is distinguished from public sources in the
statewide tabulation. Domestic water supply is usually determined by applying

ESTIMATED USE OF WATER IN THE UNITED STATES -- MANDATORY ELEMENTS

2000 Data Collection Form

State

Data Element	Public Supply	Domestic		Total Population (000s)
Population served - Total (000s)		(calculated)		
Withdrawals - GW - fresh				
Withdrawals - SW - fresh				

County

Name: _____   County Code: _____

Data Element	Public Supply	Industrial	Thermoelectric Once-through	Other	Mining	Livestock	Aquaculture	Irrigation
Withdrawals - GW - fresh					†	*		
Withdrawals - GW - saline					†			
Withdrawals - SW - fresh					†	*	‡	
Withdrawals - SW - saline					†			
Acres Irrigated - sprinkler (000s)								
Acres Irrigated - micro (000s)								
Acres Irrigated - surface (000s)								

† TX, MN, FL, OK, PA, CA, UT, AZ, NE, AK, IN, WY

* CA, TX, OK, NC, NE, IA, KS, MO, WI, MN

‡ ID, MS, LA, AR, NC, CA, AL, UT

Aquifer

Name: _____   Aquifer Code: _____

Data Element	Public Supply	Industrial	Irrigation
Withdrawals - GW - fresh			

FIGURE 3.2 USGS water use data collection form for year 2000. Form shows only those elements that are mandatory. One form is filled out for each county. Black elements are not used. SOURCE: USGS NWUIP.

estimation methods, such as coefficients for individual water use, rather than by tabulating measured values (USGS, 2000).

Industrial water is water used to manufacture products such as steel, chemicals, and paper and water used in petroleum and metals refining. Industrial water use often is locally dominated by only a few large industries. Data on industrial water use often are gathered from permitting information and customer records (USGS, 2000). *Thermoelectric water* is water used in the generation of electric power in fossil, nuclear, biomass, solid waste, and geothermal plants. The major sources of water use data for power plants are the plants themselves and allocation data from regulatory agencies. *Irrigation water* is water used to cultivate crop plants; it is by far the largest water use in the western United States. Data on irrigation water use come from many sources, including direct measurement and regulatory allocated amounts. *Aquacultural* water is water used to commercially farm fish and shellfish; the major sources of data are the same as those for irrigation. *Mining water* is water used during the extraction and processing of minerals and hydrocarbons. Water use data for mining often are obtained from agencies that regulate the mining industry. Finally, *livestock water* is water used for farm animals. Water use in this category often is determined by census and empirical coefficients for per capita animal use per unit time (USGS, 2000).

METADATA AND UNCERTAINTY

In an ideal world, all water use data would be obtained by direct measurements. Irrigation water and industrial water could be monitored by gages and other instruments, much as many public water supplies measure the volumes of water they extract from streams and wells. This kind of monitoring, even if legally mandated by local and state governments, would be very costly. In arid regions of the nation, water is a critical material commodity, and ignorance of real water use sometimes becomes politically expedient with respect to permitted or negotiated water allocations. In contrast, there can be little political incentive to monitor and regulate water use in water "rich" states such as those of the Northeast, given the perception of an almost unlimited supply in such states.

There has been little scientific evaluation of the quality of tabulated water use data provided to the USGS by state and other agencies for the national water use report that have been published every five years by the USGS. Qualitatively, water use data for public water supply and in-stream water (thermoelectric) use are probably fairly accurate where water withdrawals are directly measured at the pump or are estimated by power usage. Domestic and agricultural water use data, in comparison, are generally poorer in quality because they are often estimated by coefficients or maximum permitted amounts of water, rather than by direct measurement. (An exception is some major western irrigation projects, where diversions, surface water deliveries, and/or groundwater pumped may be measured or closely estimated to satisfy legal requirements.) Industrial water use is probably least certain because of the large variability of water use in industrial applications, even with a single industrial class. That water use data throughout the country are substantially different regionally and locally adds to the general uncertainty, compounded by nonreporting of *metadata* (i.e., data and information about data quality) on the reliability of methods used to estimate water use. An introduction to metadata is given in Box 3.3.

Water use data, in contrast to many other kinds of data, are usually presented in tabular or statistical form, sometimes aggregated by watersheds or politically defined units such as county, state, and nation. Although it is understood that many water use data are obtained by extrapolation or inference (e.g., using various water use coefficients), there have been few studies that directly determine how much error is embedded in published water use maps and aggregated estimates. Without such error analysis, valid interpretation of the water use data can be seriously compromised. For example, has the amount of water used annually

---

BOX 3.3
Metadata

Simply defined, "metadata" means "data about data." Metadata document how data are obtained and their accuracy. Metadata enhance the usefulness of data by assigning statistical or qualitative accuracy and precision to data. Geographic information system approaches have formalized how metadata are prepared for many geographic datasets. Some metadata standards (e.g., the Federal Geographic Data Committee (FGDC) standard) have hundreds of different elements describing data. Metadata are the background information that describes the content, quality, condition, and other appropriate characteristics of geospatial data.

Metadata have been used for many years when special information is presented. For example, the legends of paper maps contain metadata. As legends, metadata are clear and easily understood by map users. When map data are digital, metadata are equally important, but developing, using, and maintaining such metadata require more attention by data producers and map users who may need to modify data to suit their particular needs.

Metadata are explicitly used in extensive geographic information systems. Metadata also are used, but less formally, in all scientific studies. For example, the "methods" sections in scientific papers on water resources almost always contain estimates or calculations of errors in sampling, chemical, hydrologic, and other basic data. Confidence intervals reflecting uncertainty (such as the standard deviation about a statistical mean) are sometimes shown on graphs with the data points. Sometimes, *ranges* of values instead of discrete contour lines are used to indicate data uncertainty in two-dimensional representations of continuous data distributions, such as chemical concentrations. Describing the uncertainty of scientific measurements or estimates is an explicit part of the scientific method, regardless of whether such description is formally called "metadata" or not.

in the United States really remained about the same during the past decade, despite increasing population? Could this widely reported conclusion actually be caused by errors in regional water use estimates that, when combined with shifting regional populations, result in an apparent national stagnation in water use rather than in increasing use?

Water use experts have had to resort to estimation methods for many of the water withdrawal classes because of the legal, financial, and political constraints that limit getting hard data. For example, domestic water withdrawals and livestock water use are commonly estimated by multiplying population figures by coefficients. Irrigation water withdrawal is often estimated by multiplying acreage by assumed water needs of the crop rather than by measuring actual water pumped and applied. One of the best descriptions of how such estimates are made is found in Snavely (1986), which details the water use data collection programs and regional database of the Great Lakes-St. Lawrence River basin states. This report is enlightening because it shows how broad the range of estimation coefficients for water use can be within a geographic area with similar water availability. Often, the coefficients used for agricultural and domestic use vary by a factor of 10. Different states in the basin use very different protocols to determine how and when to tabulate water use. The result is a patchwork of disparate databases that the USGS must attempt to synthesize into a meaningful whole. Compiling such a patchwork into a meaningful and scientifically defendable whole is probably not possible, given the apparent large variability in both estimation and tabulation methods even within a similar climatic region.

The USGS needs to evaluate the scientific merits behind the coefficients and the estimation methods used to aggregate water use in the nation from first principles and primary literature and tie variations of the coefficients to climatic or other factors that can be deterministically related to water use intensity. Where coefficient data are not scientifically robust, the USGS should invest in studies that statistically sample and determine how water is used in order to develop its own coefficients that can be used to determine water use at the national scale, leaving local water use determinations to local governmental units. This does not mean that the committee does not support the collection of high-quality water use data. Rather, it acknowledges the reality that the states place varying degrees of emphasis on water use data collection and that the USGS has limited influence on state law and policy.

Thus, the USGS should not view water use compilation as a "census" of additive parts as it has done in the past, but rather, it should look at water use as a science and investigate it using appropriate scientific approaches. The USGS in its National Water Quality Assessment Program selected a set of discrete sampling regions in representative river systems to derive the status of the nation's surface and groundwater quality. It did not try to sample all rivers and waters in the nation. Rather, it is developing scientific methods to apply what is learned from well-studied regions to areas with less available data. Likewise, the stream gaging program researches methods to estimate flows in ungaged streams. The USGS should in similar fashion find ways to use information gained from the excellent water use datasets that exist, or can be easily created, to make estimates in data-poor areas.

VALUE OF DATA AT VARIOUS SCALES OF SPACE AND TIME

The USGS aggregates water use data into a smaller number of categories from state and smaller political units (e.g., county) to produce statewide and national-level compilations. As appropriate to such a compilation, these data are most useful for characterizing regional water use trends and national changes in water use. However, without systematic scientific studies testing the reliability of the compiled data, the usefulness of the compilations is questionable.

The data richness at more detailed scales not reported in the national synthesis, however, should not be underestimated. For example, data on the increase in agricultural acreage from 1992 to 1997 (Figure 3.3) provide a national snapshot of potential changing water use that is not—but arguably should be—captured in the USGS water use reports.

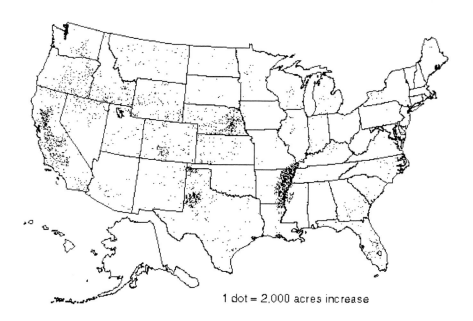

1 dot = 2,000 acres increase

FIGURE 3.3 Areas of increase in irrigated acreage in the United States, 1992-1997. SOURCE: Modified from USDA (1999a).

Water use trends and use patterns can be evaluated when excellent site-specific water use data are available—particularly physical measurements of water use at major point sources. The cast study in the next section shows how excellent water use data in Arkansas enabled the committee to explore major water use issues. It demonstrates that meaningful and accurate water use databases have the same intrinsic value as spatial datasets commonly collected as part of national resource evaluation and initiatives (e.g., streamflow, meteorological data, soils characterization, etc.).

STATE-LEVEL WATER USE DATA: THE ARKANSAS CASE STUDY

One of the most impressive state water use programs reviewed by the committee is that carried out in Arkansas by the USGS in collaboration with the Arkansas Soil and Water Conservation Commission. Initiated some 20 years ago, the program today has matured into a comprehensive inventory of monthly data from more than 40,000 surface and groundwater withdrawal points. The program also plays a vital role in the state's effort to manage water resources in heavily pumped regions, especially the Mississippi Alluvium aquifer in eastern Arkansas. The summary of water use in Arkansas presented here is developed from data provided courtesy of the USGS Arkansas District Office.

Water Use Data in Arkansas

Water use data for both surface and groundwater are collected to meet the requirements of state law in Arkansas (Act 81 of 1957 as amended and Act 1051 of 1985). Users who withdraw at least 50,000 gallons per day (groundwater) or 1 acre-foot per year (surface water) must report their monthly withdrawals each year to the Arkansas Soil and Water Conservation Commission (ASWCC). They are not required to report their consumptive use. During the early 1980s, the ASWCC received approximately 10,000 water use registrations annually and stored the data on the ASWCC computer. In 1983, the USGS entered into a cooperative agreement with the ASWCC to support this data collection program by providing a Site-Specific Water Use Data System (SWUDS) and by providing support and training to ASWCC personnel. In 1985, USGS personnel began entering water use data and providing quality assurance. The USGS also printed water use registration forms, reminder cards, and certificates for ASWCC.

During the 1990s, the program evolved with the development of customized reporting and software specifically configured for Arkansas. In 1993, the USGS produced 29 county and area reports of site-specific water use in Arkansas. In 1994, the USGS began installing custom water use software (WUDBS) in 30 County Conservation District offices in Arkansas (on NRCS computers) for remote entry of water use data. Since 1994, the USGS has trained and provided computer software support to Conservation District personnel in 30 of the 75 counties in Arkansas. The USGS currently enters water use data and prints forms, reminder cards, and certificates for all nonagricultural water use categories, and it conducts the water use survey for the other 45 counties in the state (approximately 10,000 measurement points).

In 2001, the USGS began the process of migrating the water use software and database to the Internet, enabling remote entry of data using a web browser (Figure 3.4). When complete, the web database and software will enable the USGS to maintain one centralized database rather than 31 separate databases. Also, the web interface provides the USGS and ASWCC with a quick, secure means of accessing the most current water use data in Arkansas.

Overview of Water Use in Arkansas

Arkansas, with a population of 2.6 million in 1999, uses water at a mean annual rate of 7,000 million gallons per day (MGD) (7.8 million acre-feet per

FIGURE 3.4 Web interface to Arkansas water use database system (secured site; login and password required). SOURCE: USGS Arkansas District Office.

year). The largest public water supply systems are associated with the principal cities: Little Rock, Jonesboro, Fayetteville, Fort Smith, and Pine Bluff (Figure 3.5). In most of the more rural counties, public water supply averages less than 5 MGD. However, total water use is dominated by withdrawals for irrigation, focused in the eastern counties of the state adjacent to the Mississippi River, where water use in some counties exceeds 500 MGD. Water use in Pope County is very high arising from the Arkansas Nuclear One power plant, located on the Arkansas River.

The trends in water use, shown in Figure 3.6, show the dominance of irrigation over public supply and also demonstrate that irrigation water withdrawals grew by approximately 50 percent from 1980 to 1995.

Water use in White County is about average for Arkansas counties, totaling approximately 114 MGD. As shown in Figure 3.7, this figure is dominated by irrigation, with much smaller amounts for public supply and other agricultural water use. Detailed information about crop and aquaculture water use for White County is shown in Table 3.1. The first two columns of the table show the number of reported applications of water and the number of acres to which the water was applied. The remaining columns show statistics of the water use per acre based on the data reported to the ASWCC.   For example, for berries, 50 acres are irrigated with an average of 4 acre-feet per acre of water, so the total water use is 200 acre-feet.

The Arkansas water use database contains data for approximately 44,700 groundwater and surface water pumping withdrawal points. Of these, approximately 5,600 points are from surface water and 39,100 are from groundwater. Thus, one-eighth of the water withdrawal points are from surface water and seven-eighths from groundwater. As shown in Figure 3.8, both surface and groundwater withdrawals are concentrated in eastern Arkansas.

Spatial Patterns in Surface Water Use

When viewed at a distance, there is no particular spatial pattern in the surface water use points, but a closer view in Figure 3.9 reveals that the surface water use points are clustered along the principal river systems in eastern Arkansas. Indeed, a more detailed view would find that most of these points are located right on these river systems, and they can be locationally referenced with river addresses, as shown later in this chapter.

Spatial Patterns in Groundwater Use

The principal aquifers underlying Arkansas are the Ozark Plateau in the north, the Mississippi River alluvium underlain by the Mississippi Embayment in the east, and a small portion of the Edwards-Trinity aquifer in the west. Of these,

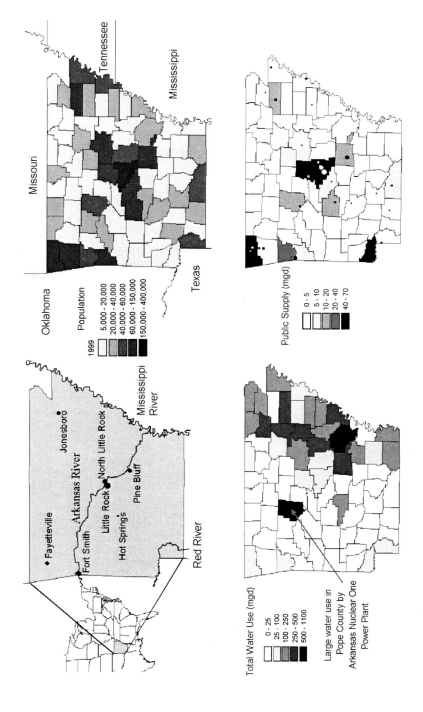

FIGURE 3.5 Population and water use in Arkansas. Data source: USGS Arkansas District Office.

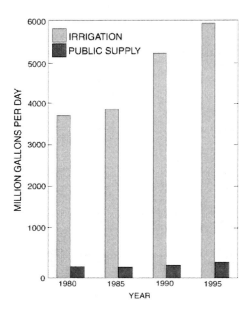

FIGURE 3.6 Trends in irrigation and public water supply for Arkansas from 1980 to 1995. Data source: USGS Arkansas District Office.

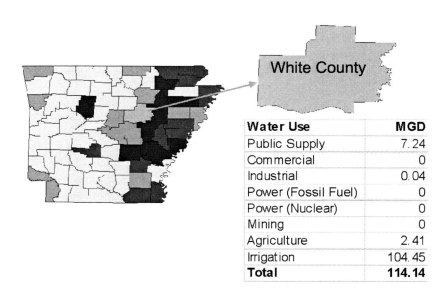

Water Use	MGD
Public Supply	7.24
Commercial	0
Industrial	0.04
Power (Fossil Fuel)	0
Power (Nuclear)	0
Mining	0
Agriculture	2.41
Irrigation	104.45
Total	**114.14**

FIGURE 3.7 Water use in White County, Arkansas in 1997, expressed as million gallons per day. See Figure 3.5 for map legend. Data source: USGS Arkansas District Office.

TABLE 3.1 Agricultural Water Use in White County, Arkansas, 1995

Crop	Number of reported applications	Sum of acres	Average	Minimum	25th percentile	50th percentile	75th percentile	Maximum	Variance	Standard deviation	Total applied (Acre-feet)
IDLE	406	0.00	0.00	0.00	0.00	0.00	0.00	0.00	0.00	0.00	0.00
RICE	684	29827.81	2.99	2.50	3.00	3.00	3.00	5.00	0.02	0.13	89404.23
CORN	4	399.00	0.62	0.50	0.50	0.50	0.75	1.00	0.06	0.25	231.50
SOYBEANS	487	28350.30	0.52	0.32	0.50	0.50	0.50	1.00	0.01	0.10	14739.80
MILO	24	2506.00	0.52	0.50	0.50	0.50	0.50	1.00	0.01	0.10	1301.00
HAY	4	195.00	1.06	0.50	0.75	1.00	1.38	1.75	0.27	0.52	222.50
VEGETABLES	2	35.00	0.43	0.43	***	0.43	***	0.43	0.00	0.00	15.00
BERRIES	1	50.00	4.00	4.00	***	***	***	4.00	***	***	200.00
GRAPES	1	30.00	0.50	0.50	***	***	***	0.50	***	***	15.00
FRUIT TREES	4	155.00	1.84	0.36	0.43	0.50	3.25	6.00	7.70	2.77	621.90
SOD	11	1002.00	6.55	1.00	1.00	7.00	12.00	12.00	30.27	5.50	10854.00
BEEF CATTLE	1	100.00	0.28	0.28	***	***	***	0.28	***	***	27.60
ANIMAL AQUACULTURE	38	721.00	3.46	0.80	2.00	3.00	4.00	7.00	3.97	1.99	2679.50
MINNOWS	1	7.00	1.25	1.25	***	***	***	1.25	***	***	8.75
DUCKS	2	300.00	1.00	1.00	***	1.00	***	1.00	0.00	0.00	300.00
Sports & recreation cl	2	96.00	1.00	1.00	***	1.00	***	1.00	0.00	0.00	96.00
Amusement & Recreation	1	210.00	0.50	0.50	***	***	***	0.50	***	***	105.00
Totals	0	0.00	***	***	***	***	***	***	***	***	0.00

SOURCE: USGS Arkansas District Office. 1,120 acre-feet/year = 1 MGD

Surface water: 5,600 points   Groundwater: 39,100 points

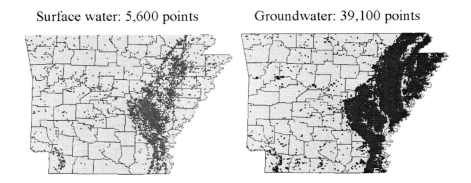

FIGURE 3.8 Surface and groundwater withdrawal points in Arkansas (1997).   Data source: USGS Arkansas District Office.

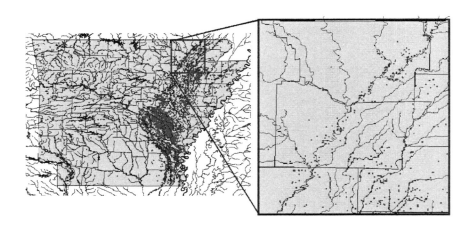

FIGURE 3.9 Spatial patterns of surface water use along rivers in Arkansas. Data source: USGS Arkansas District Office.

by far the most important water supply is provided by the Mississippi Alluvium, which supports 33,700 groundwater wells or 86 percent of all the wells in the state. The spatial patterns of groundwater withdrawal points evident in Figure 3.8 are shown in Figure 3.10 to be closely associated with the boundaries of the Mississippi Alluvium.

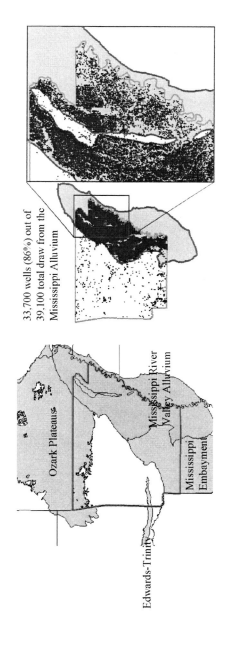

FIGURE 3.10 Spatial patterns of groundwater use, especially the Mississippi Alluvium.
Source of aquifer information: USGS (1970). Source of well information: USGS Arkansas District Office.

Temporal Patterns of Water Use

As part of the USGS Cooperative Water (Coop) Program, each year, personnel in the county offices of the Arkansas Soil and Water Conservation Commission collect monthly water use estimates for each water withdrawal point. The USGS archives and organizes this information into a statewide database. Table 3.2 summarizes statistics of irrigation and public supply water use, while Figure 3.11 illustrates the time patterns of these uses.

As shown in Table 3.2, there are about seven times as many water withdrawal points from groundwater as compared to surface water, for both irrigation and public supply. The average amount of water withdrawn from an individual groundwater well is about 12 ac-ft/mo. for both irrigation and public supply, while for surface water, the average amount withdrawn at a single withdrawal point is larger, about 21 ac-ft/mo. for irrigation and 234 ac-ft/mo. for public supply. The net result is that irrigation supply is principally drawn from groundwater and public supply from surface water, with the irrigation supply being spatially dispersed over tens of thousands of wells, while the public supply is more concentrated at hundreds of withdrawal points.

As shown in Figure 3.11, there is a seasonal pattern to both irrigation and public supply. Rates of withdrawal for the irrigation and public supply are highest in the summer months and lower in the winter, a tendency that is much more pronounced in the case of irrigation. These data demonstrate that the practice used by the USGS of sampling and tabulating groundwater and surface water use separately for each usage category is justified. If the source of water supply is not considered, important information is lost even though the total water use in any category lumps together the use from surface and groundwater sources.

Water Use Intensity

A general indicator of the stress that water use is placing on the existing natural water systems is the *water use intensity*, expressed here as the amount of water use per unit area for a county, an aquifer, or a watershed. Because water

TABLE 3.2 Numbers of Users and Average Withdrawals Per User for Irrigation and Public Supply, Arkansas

Water Source	Irrigation		Public Supply	
	No. of Users	Avg. Withdrawal Per User (ac-ft/mo.)	No. of Users	Avg. Withdrawal Per User (ac-ft/mo.)
Groundwater	36,053	12.7	894	11.6
Surface Water	5,049	21.1	132	234

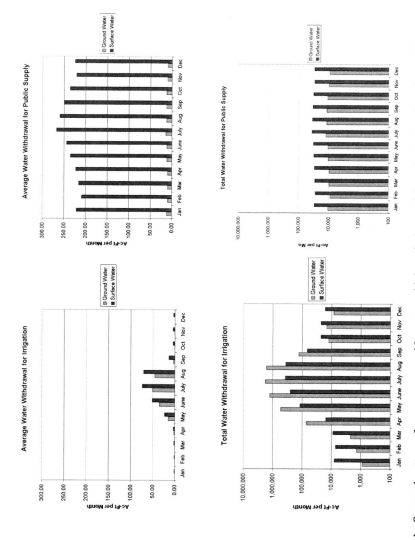

FIGURE 3.11 Seasonal patterns for average monthly water withdrawals (per withdrawal point) and total monthly water withdrawals in Arkansas in 1997, for irrigation and public water supply. Data source: USGS Arkansas District Office.

use is expressed in dimensional units of $[L^3/T]$ (where L is length and T is time) and area in units of $[L^2]$, it follows that the water use intensity is expressed in units of $[L^3/T] / [L^2] = [L/T]$. It can thus be calculated in inches per month or inches per year and compared to precipitation or runoff expressed in comparable units.

Figure 3.12 shows the resulting spatial patterns of water use intensity for surface water, groundwater, and total water use in Arkansas. In the calculation of the surface water use intensity, water use for power production was excluded because it occurs in only a relatively few locations and because it is of such a large magnitude at those locations that it dominates by far all other categories of water use. Total water use intensity averages 2.9 inches per year over all counties and has a maximum value of 17.3 inches per year in Arkansas County. Surface water use intensity averages 0.8 inches per year over all counties and has a maximum value of 8.2 inches per year in Arkansas County. Groundwater use intensity averages 2.1 inches per year over all counties and has a maximum value

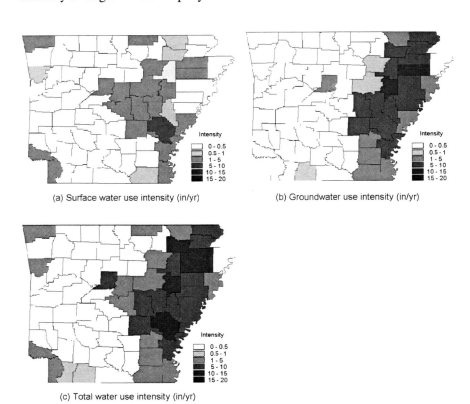

(a) Surface water use intensity (in/yr)

(b) Groundwater use intensity (in/yr)

(c) Total water use intensity (in/yr)

FIGURE 3.12 Water use intensity in inches per year for (a) surface water, (b) groundwater, and (c) total water use (surface plus groundwater). Data source: USGS Arkansas District Office.

of 12.3 inches per year in Woodruff County. Thus, excluding water use for power production, about 70 percent of the total water use per unit of land area in Arkansas is supported by groundwater and about 30 percent by surface water.

Water Stress

The magnitude of the water use intensity in inches per year can be compared with the water fluxes through the components of the hydrologic cycle, also measured in inches per year. A convenient benchmark for this comparison is the spatial distribution of annual precipitation. A *water stress index* can be defined as the ratio of the water use intensity to the precipitation rate. As shown in Figure 3.13, the precipitation in Arkansas ranges between 44 and 56 inches per year,

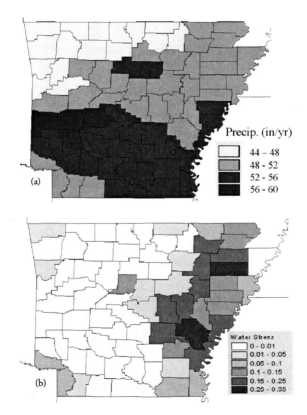

FIGURE 3.13 (a) Average annual precipitation, 1961-1990 and (b) water stress index (ratio of total water use intensity to annual precipitation). Precipitation data source: USDA Climate Maps and Digital Data (http://www.ftw.nrcs.usda.gov/prism/prism.html). Water use data source: USGS Arkansas District Office.

with less precipitation in the north than the south and with the highest precipitation in the hills of southwestern Arkansas. The water stress index map, shown in Figure 3.13, has an average value of 0.06 over the state and a maximum value of 0.32 in Arkansas County. Thus, approximately 6 percent of annual precipitation is withdrawn from surface and groundwater resources on average over Arkansas, and 32 percent or nearly a third of the annual precipitation is pumped out in Arkansas County. As before, these figures do not include water use for power production. Although these kinds of calculations may be misleading at a local level because of additional sources of water (e.g., Mississippi River via recharge through the alluvium), they are likely to have considerable application at a regional or national scale.

If only consumptive water use rather than total water use were used, these water use intensities and stresses would be somewhat lower. Consumptive use for irrigation (the most important water use category) in Arkansas in 1995 was estimated by Holland (1995) to be 4,393 MGD of the 5,936 MGD withdrawn, or 74 percent. The remainder, approximately 26 percent, of the water withdrawn was returned to either groundwater or surface water and was available for reuse if it was of acceptable quality.

Linear Referencing of Water Use Points to River Locations

The USGS has been a leader in displaying water data on the Internet for map information on both water features and water measurements. The main surface water features useful for map description are watersheds and river reaches. If desired, water use points can be located on river reaches within watersheds using a river addressing system.

Hydrologic Cataloging Units

The surface drainage system of the United States has been divided into a hierarchical system of hydrologic cataloging units. At the top of the hierarchy are two-digit water resource regions, of which there are 18 covering the continental United States. Region 11 describes the Arkansas and Red River basins, and Region 8 the Lower Mississippi basin. The regions are further subdivided into four-, six- and eight-digit cataloging units as shown in Figure 3.14 for Regions 8 and 11.

The eight-digit cataloging units are the most widely used drainage areas for cataloging hydrologic data in the United States. There are 2,156 such units in the continental United States, of a size roughly equivalent to counties. Each eight-digit unit has a hydrologic unit code. For example, cataloging unit 08020402 is the Bayou Meto unit in northeastern Arkansas. It is subbasin 2 of Basin 4 within Subregion into 2 of Region 8 in the cataloging unit hierarchy.

FIGURE 3.14 Hydrologic cataloging units of the United States. The smallest units shown in the figure are eight-digit cataloging units. SOURCE: USGS (http://water.usgs.gov/GIS/huc.html for description of HUCs and http://water.usgs.gov/GIS/huc.html for data retrieval).

RIVER REACHES

The USGS and the U.S. Environmental Protection Agency (EPA) have together developed the National Hydrography Dataset (http://nhd.usgs.gov), a digital description of the rivers, streams, and waterbodies of the United States (see Figure 3.15). Developed nationally at a scale of 1:100,000, the National Hydrography Dataset is progressively being improved by adding high-resolution 1:24,000-scale hydrography data in some states. The river and stream network of this dataset has been separated into reaches and labeled using a Reach Code, which is a 14-digit identifier, unique throughout the United States. For example, Reach 08020402000077 is river segment 77 within cataloging unit 08020402.

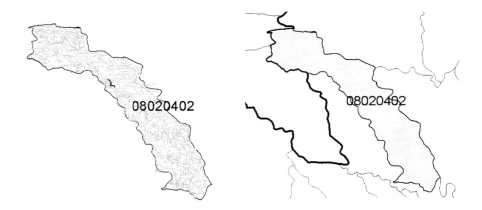

FIGURE 3.15 The National Hydrography Dataset river and stream network for eight-digit hydrologic cataloging unit 08020402, named Bayou Meto. SOURCE: USGS National Hydrography Dataset (http://nhd.usgs.gov).

River Addressing

Because each river reach is a uniquely labeled line, an "address" location can be defined at any point on that line as the *measure*, or percent distance from the downstream end of the line. For example, location A, in Figure 3.16 can be specified as having an address of 0.38 (i.e., 38%) on river segment 77 in cataloging unit 08020402. This river addressing system is similar to that normally used for street addressing. For example, 384 Vine Street, Austin, Texas 78712, indicates a unique address *number* (384) on a *line* (Vine Street), within an *area* (zip code 78712). Similarly, point A is at *number* 0.38 on a river reach *line* within a cataloging unit drainage *area*. The use of percent distance in the National Hydrography Dataset as the measure location is replaced in other river addressing systems by linear measures such as river miles or kilometers from the river mouth.

Linear Referencing of Water Use Points

Linear referencing is a process of locating information by addresses on lines. This concept, most commonly used for mail addressing, is being used to take point data whose latitude and longitude are known and locate those points on river systems. Figure 3.17 shows four surface water use points on the Bayou Meto. Each water use point has an identifying number, a use type (IR is for irrigation, AG is for general agriculture), a source (SW = surface water), an

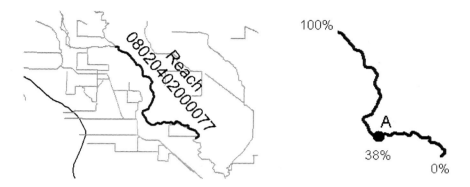

FIGURE 3.16 Location on a river reach, given by percent distance from the downstream end of the line.

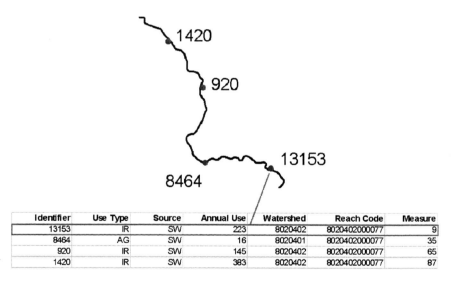

Identifier	Use Type	Source	Annual Use	Watershed	Reach Code	Measure
13153	IR	SW	223	8020402	8020402000077	9
8464	AG	SW	16	8020401	8020402000077	35
920	IR	SW	145	8020402	8020402000077	65
1420	IR	SW	383	8020402	8020402000077	87

FIGURE 3.17 Example of linear referencing of surface water use points using an event table. Note the one-to-one correlation of lines in the table and water use points on the map.

annual water use, a watershed number, a reach code, and a measure value. Each row in the table shown is a linear *point event*, which means that its location is known not by its latitude and longitude but rather by its location on the river reach line.

Once all of the surface water use points are similarly located, questions such as the following can be asked: What is the water use upstream or downstream of this location? How does this compare with existing flows in this river? If wastewater discharge points are similarly located, other questions can be asked: What is the flow distance between this water use point and the nearest upstream wastewater discharge? If a spill occurs upstream, which downstream water users will be affected?

The association of water use sites and the National Hydrography Dataset creates a powerful combination of information capable of answering water management questions that cannot otherwise be posed. And the key question in this association is not so much *how much* water is used, but *where* it is withdrawn from surface water sources. This location information is something that can be accurately measured and the quality controlled using global positioning systems. Indeed, the USGS in Arkansas has undertaken an extensive survey to perform quality control on the location data for the water use data points in its database.

When all of the surface water use points in the Arkansas database are displayed over the National Hydrography Dataset river reaches, as shown conceptually in Figure 3.18, most of the points fall on the mapped reaches, but there are some points that do not. These are points where water is being diverted from smaller rivers and streams than are depicted on 1:100,000-scale maps. If these 1:100,000-scale data were replaced by more detailed 1:24,000-scale data, most of the remaining points would be located on a mapped river reach.

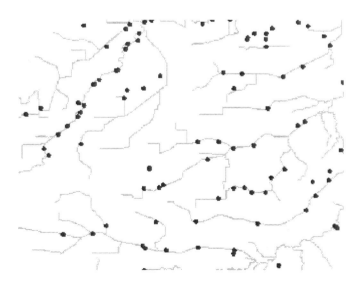

FIGURE 3.18 Water use points and National Hydrography Dataset river reaches in a hypothetical Arkansas basin.

NATIONAL WATER USE DATA:
FEDERAL SITE-SPECIFIC DATABASES

The Arkansas case study illustrates the richness of the information contained in a comprehensive, statewide, site-specific database and some of the many ways in which this information can be applied. However, the survey of state water use data collection programs presented in Chapter 2 and Appendix A shows that such databases exist in only about 20 states. Fortunately, many classes of water use facilities are federally regulated, so information bases exist at the national level as well. Although these datasets are not all equal in quality, they share the critical element of providing consistent baseline information from state to state. The potential importance of these databases becomes especially clear in Chapter 7, where we discuss the development of a national dataset of water use information.

In the following sections, four federal datasets are profiled, three of which are site-specific. These datasets are relevant to the estimation of water use in the areas of public water supply, power plants, wastewater, and agriculture.

Public Water Supply Database:
Safe Drinking Water Information System (SDWIS)

The EPA has compiled a Safe Drinking Water Information System (SDWIS) (HtmlResAnchor http://www.epa.gov/safewater/sdwisfed/sdwis.htm) to support the Safe Drinking Water Act. This database covers all public water supply systems in the United States serving at least 25 people or 15 connections on a year-round basis. The USGS supports the EPA by providing quality control information on the latitude-longitude coordinates of the approximately 6,600 surface water intake locations and 104,000 groundwater withdrawal locations documented in the SDWIS database.

Energy Generation Facilities Databases:
Energy Information Administration

The Energy Information Administration (EIA) (http://www.eia.doe.gov/) has extensive databases on locations and characteristics of energy generation facilities in the United States. This information includes monthly time series on electricity production at power plants. These consistent national data enhance and complement the information currently used to estimate water use for power generation by nuclear and fossil fuel thermoelectric power plants.

Wastewater Discharge Database: EPA Permit Compliance System

The EPA maintains a Permit Compliance System to document information on facilities that have permits to discharge wastewater into rivers. This system contains information on the life of the permit, the permitted discharge, and discharge monitoring data (http://www.epa.gov/enviro/). Figure 3.19 shows the locations of wastewater-discharge site locations stored in the Permit Compliance System. Wastewater discharge data may also be useful in estimating water use when they are combined with an understanding of the active industrial processes. One source of such information is a series of so-called Industry Sector Notebooks, published by the EPA Office of Compliance. The sector notebooks, published for over 30 major industrial sectors, offer a holistic "whole-facility" approach to industry-specific manufacturing processes and to pollution issues for specific industrial sectors. The Permit Compliance System together with the sector notebooks can be used to estimate industrial water use may be estimated from the national database of permitted wastewater dischargers.

This methodology would have limitations. These sector notebooks are incomplete, fail to account for recycled water, and do not account for water embodied in different types of processes. Further, the databases themselves are incomplete for industries located where permits are not granted.

Agricultural Databases:
Census of Agriculture and National Resource Inventory

The only large category of water withdrawal that is not federally regulated at each individual water use site is irrigated agriculture. Several large national databases help to partially address this issue. Every five years, the U.S. Department of Agriculture (USDA) compiles a comprehensive survey of national agricultural activities called the Census of Agriculture (http://www.nass.usda.gov/census/). As part of the Census, irrigated acreage is estimated for every county in the United States, as suggested in Figure 3.20. The Farm and Ranch Irrigation Survey (FRIS) (http://www.nass.usda.gov/census/census97/fris/fris.htm), conducted the year following the Census of Agriculture, provides supplemental data on irrigation water use by source (groundwater, on-farm surface water, and off-farm), farm water applications (*not* withdrawals or consumptive use), and irrigation practices. The stratified random sample design of the FRIS provides rigorous confidence limits on all estimated quantities. Another USDA survey is the National Resource Inventory, which is a statistically based sample of land use and natural resource conditions and trends on U.S. nonfederal lands (http://www.nhq.nrcs.usda.gov/NRI/1997/). A major limitation is that neither of these programs reports site-specific data because of confidentiality requirements. Despite this, agricultural water use represents a natural area for increased cooperation between the USGS and USDA. Even aggregated data may serve as a check for a site-specific survey, which the USGS may choose to carry out. Fertile areas for this interagency collaboration include estimating agricultural water use and determining the effect of this use on the quantity, quality, and ecology of the nation's water resources. The Survey's expertise and interest in linking water use

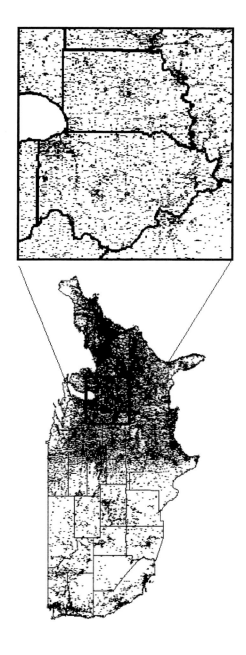

FIGURE 3.19 Locations of wastewater discharge points in the EPA Permit Compliance System.
Data source: Thomas Dabolt, EPA, written communication, 2001.

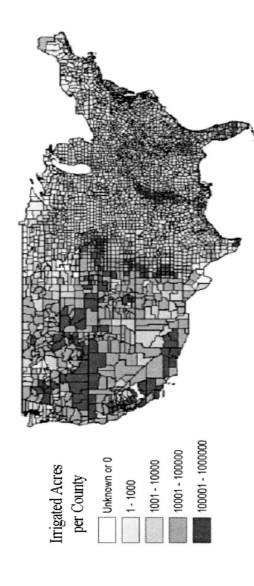

FIGURE 3.20 Acres of irrigated land in the United States by county in 1997.
SOURCE: 1997 Census of Agriculture, as reported in National Atlas of the United States (http://www.nationalatlas.gov/agcensusm.html).

to the source stream or aquifer complements the USDA emphasis on economic inputs and characteristics of the nation's agricultural production.

In chapters 4–6 the report moves beyond water use *data* to water use *estimation*. Chapter 4 begins with a summary of the many different estimation techniques that may be used in the field of water use. Two techniques are then investigated in greater detail: direct estimation through stratified (and unstratified) random sampling (Chapter 5) and indirect estimation using linear regression models that explain water use based on corresponding demographic, economic, and climatic data (Chapter 6).

CONCLUSIONS AND RECOMMENDATIONS

Good-quality water use data are critical for many purposes, including (1) ensuring adequate availability of water as future water demands fluctuate due to changes in climate, urban growth patterns, agricultural practices, and energy needs and (2) settling interstate and intrastate conflicts over water resources. Given the widespread economic and political implications of water use compilations, *it is very important that water use data be compiled with metadata defining the uncertainty in the numbers.* Metadata document how data are obtained and enhance the usefulness of data by assigning statistical or qualitative accuracy and precision to data. The USGS aggregates water use data to produce statewide and national-level compilations. Without systematic scientific studies testing the reliability of the compiled data, the usefulness of the compilations is questionable.

At present, because there is only a limited national-level quality assurance/ quality control program for most of the received data, it is difficult to ascribe confidence intervals to the data. *Fundamentally, water use needs to be treated with the same scientific rigor used to quantify other aspects of the water budget,* even if water use data inherently have much greater uncertainty.

Because there now is great uncertainty in many estimates of water use, *it is important to develop the scientific tools to address and minimize the uncertainty to the extent possible.*

Finally, high-quality, site-specific datasets open up many ways of visualizing water use. The Arkansas case study developed in this chapter demonstrates how these data can be used to show spatial and temporal patterns in groundwater and surface water use, water use intensity, and water stress. *The USGS should encourage the development of site-specific databases where none exist and should work with cooperators to fully exploit the power of these databases.*

4

Overview of Water Use Estimation

Though conceptually appealing, it is impossible to fully account in practice for all the individual decisions and behaviors that constitute the nation's water use. Nevertheless, statistical sampling and indirect methods can be used to estimate aggregate water use. The diverse goals of water use estimation, and the varied categories and methods used to estimate water use, suggest no single approach can be—or should be—expected to satisfy all the requirements of the National Water-Use Information Program (NWUIP). The challenge of identifying an appropriate methodology—extensive data collection or indirect estimation—is evident in the National Handbook of Recommended Methods for Water Data Acquisition (USGS, 2000):

> Water use can be determined either for site-specific facilities or for categories of water use for a given area. Determining site-specific water use involves water withdrawal, delivery, release, and return-flow data. Area estimates are based on coefficients relating water use to another characteristic, such as number of employees, and applying them to an inventory of site-specific users or by measuring a statistical sample of the user population.

> Choosing whether to use site-specific data or area estimates to estimate water use will depend on the objective of the water use data collection; availability of statewide reported data; availability of time, manpower, and funds; and the area to be covered. The objective of the water use data collection determines the required degree of accuracy and reliability and identifies individual data elements that are relevant.

In contrast with *data collection* to compile a complete inventory of national water use, water use *estimation* requires statistical inference using inherently mixed-quality data that are inconsistently sampled in time and space. Significant characteristics of water use estimation in the NWUIP include the following:

• Estimates are necessarily grounded in local, individual data records. Estimates require integrating data of mixed quality that are collected by other agencies for other purposes and that are derived from data collection protocols generally neither controlled nor modifiable by NWUIP.
• The water use database includes data on individual facilities that can be accurately tallied (such as public wells and permitted intakes), for which the associated water use is variable and uncertain and may be bounded (e.g., by a permit limit or water right).
• The water use database includes data on diffuse uses that can be identified only through indirect measures (such as number of employees, dollar value of commercial manufacturing, cultivated acreage) and that can vary dramatically with climate variability and national economic trends.
• Although grounded in local site-specific data, water use estimates must meet broader needs for information at state, regional, watershed, and national scales.

Historically, U.S. Geological Survey (USGS) national water use estimates have not been developed for a distinct client or constituency. Nor have they been developed in response to specific scientific or management goals. The result is a contextual void for evaluating the quality, accuracy, and value of the many data sources and estimation techniques used to populate this unique national database. The requirements for a water use estimation program suggested in this report underscore the need to quantify the accuracy and uncertainty of the national water use estimates.

A variety of methods will be required to populate a national water use database. The choice of estimation methods must balance the quality and availability of data, the information needs, and the available resources. The following section contrasts strengths and limitations of alternative methods for estimating water use.

DIRECT ESTIMATION

Direct estimation methods can be broadly divided into the complete inventory approach and the stratified random sampling approach. A complete inventory of water use seeks to quantify (by direct measurement or secondary records) every water use in a geopolitical or hydrologic region. The complete inventory is conceptually straightforward and has been the idealized model for the USGS water use program for several decades. It is also not possible or cost-effective in many cases, and sampling of a subset of the complete population is often the more reasonable approach.

Complete Inventory

A complete inventory presumes that all water uses can be identified and measured with sufficient resources and effort. Direct measurement reduces the water use estimation problem to a data collection and database management program. To support the conceptual model of a complete inventory of national water use, the USGS has wrestled with fundamental questions of defining water use and has developed a taxonomy of water use categories to support internally consistent accounting, comparison, and aggregation to regional and national scales. Without specific research goals or applications, the water use categories adopted to meet database challenges are intrinsically arbitrary. Indeed, the ambiguity in the purpose and goals of this data collection effort is evidenced by the evolution and redefinition of water use categories and data collections methods and even by the redefinition of water use in each of the Survey's pentannual water use assessments. Nevertheless, the USGS water use categories provide an internally consistent data structure to formulate and organize water use estimates within each state.

The complete inventory model for national water use has pragmatic limitations. Consistent information collected at monthly, weekly, or daily time intervals would be difficult to compile. Useful ancillary data, such as the price of water, crop type, and soil characteristics, are not usually included in "inventory" datasets. In practice, a truly complete inventory of every national water use is unobtainable. Nevertheless, the diligent effort to scour every source and "all available" information naturally fosters the sense that these estimates must be the "best available," regardless of the unquantified uncertainty in estimated use. Pragmatically, a complete inventory of all water uses is unachievable.

As a conceptual model, the USGS's inventory of water use by category is little changed from water use estimation techniques used over 150 years ago (see Box 4.1). The closest example to a complete inventory in the USGS water use program may be the Arkansas water use program. Here, the coincidence of a newly established statewide water use permitting program adopting the USGS water use data management technology resulted in a unique partnership by which USGS provides both technology and database management for the entire permitting system. This cooperative partnership has provided the Arkansas District with one of the most complete and current water use databases in the country. It is important to note that the Arkansas water use database is defined by the state's permit requirements and therefore emphasizes water use at the point of withdrawal. Water use information relevant for policymaking, such as domestic self-supplied water use, irrigation efficiency, the rate of industrial reuse, magnitude and location of return flows, and total stress on surface and groundwater, all require additional analysis and interpretation.

The complete inventory model is intuitively appealing and may be well suited to specific regulatory functions and permit needs. For example, river

---

BOX 4.1
Inventory of Water Use

The common use of a "complete inventory" of water-using activities, combining actual use with "water use coefficients" is neither novel nor unique to the USGS National Water Use Information Program (NWUIP). This intuitive "accounting" for water use is actually little changed (aside from the use of digital data management tools) from the methods used by the city of New York in estimating potential revenues for financing construction of the Croton Reservoir in the first half of the nineteenth century (Koeppel, 2000).

"To determine the number of potential water users and how much they might be willing to pay, the commissioners drew a broad statistical portrait of 1835 New York. From information supplied by city surveyor Edwin Smith and others, the commissioners found that below its limits on 21st Street, New York contained 30,000 houses, over 2000 back tenements, 240 boarding houses, and 40 large hotels; 2,646 taverns and 100 victualing and refectory houses; 267 bakehouses, 63 distilleries, 12 breweries, 10 porter cellars, and 70 sugar houses; 178 printing offices, about 60 silversmiths and jewelers, 58 soap and candle factories, 43 marble and stone cutting works, and 10 type foundaries; 73 hatteries and 19 curriers and morocco manufacturers; 5000 horses, 237 butchers, 100 slaughter houses, 86 livery stables, and 1 tanyard; 26 classical schools, 23 primary schools, 22 female boarding schools, 12 public schools, and 7 'African free schools'; 60 steam engines (exclusive of those at distilleries, sugar houses, and stone works), 2 gas works, and 1 chemical factory.

"Presuming that two-thirds of the houses and all other potential users would immediately sign up for Croton water, and after studying water use information solicited from other cities, the commissioners offer a table of estimated water revenue."

SOURCE: Koeppel (2000).

---

basins that are highly regulated may require not only a complete inventory of withdrawals, but of return flows as well, in order to calculate consumptive use. At a larger scale, however, the principal limitation of the complete inventory model is the practical challenge of identifying and sampling every water use relevant for a national information program. When a complete inventory is impractical, water use may be estimated directly using formal sampling techniques, such as those described in Chapter 5.

Stratified Random Sampling

The enormous variation in both the quality of available water use data and the economic and climatic patterns of water use among the states poses substantial challenges to any consistent national estimation framework. A rigorously formulated national sample design offers a practical means for developing consistent national water use estimates with uniformly quantified accuracy and reliability characteristics. One model for consistent national sample design is the Census of Agriculture of the USDA National Agricultural Statistics Service (NASS) (USDA, 1999a).

The Census of Agriculture uses a rigorous stratified sample design, employing both telephone and mail survey instruments, to develop detailed county-level estimates of national agricultural activities. Viewing the Census of Agriculture as a model for nationally consistent county-level stratified random sampling highlights a number of significant features and methodological issues that would be common to the national estimation of water use:

• The sample design rigorously emphasizes computing the standard error, and hence confidence limits, on all estimates.

• The program fosters an institutional culture of continuous improvement in sequential estimates. Each cycle of estimation draws upon lessons learned and observed trends from previous surveys, while simultaneously laying the foundation (through additional subsample survey questions) to evaluate emerging trends to be incorporated in future sample design.

• Core expertise in sampling and research design is integrated with local and regional specialists throughout the states. This structure is suggestively similar to the infrastructure of the state water use specialists working with methodological guidance and statistical expertise found in the Water Resources Division of the USGS.

The culture of a continuously improving estimation process naturally leads to methodological research and evaluation of new techniques to improve accuracy and cost effectiveness. Such techniques include the following:

• the use of screening models and follow-up surveys to evaluate the probability of membership in the sample population as part of the sampling design,

• the use of remote sensing and expert systems for crop classification, mapping, and acreage estimation,

• the use of follow-up evaluations to estimate the completeness of the target population coverage, evaluate nonresponse rates, and identify problems in order to improve future sample designs, and

• of development techniques to extend survey estimates to nonsurvey years.

Stratified random sampling can optimally allocate limited sampling resources among homogeneous subpopulations, or strata, of a well-defined statistical population. Further, this sampling approach makes no assumptions regarding the probability distribution of the water use within a stratum (i.e., a normal distribution is not assumed). If reliable population data are available, accurate direct estimates of water use can be obtained with substantially less data collection than that required for complete inventory methods. However, this reduced sampling effort must be accompanied by the additional effort required to maintain current, accurate population information. If reliable population information is not available, their consistent water use estimates can be developed using indirect estimation methods.

INDIRECT ESTIMATION

Where direct sample data are inadequate or unavailable, indirect estimation methods may be used. *Although indirect estimation methods can be extremely valuable when direct sample data are limited, these methods nevertheless require sufficient data to support calibration and verification.* The following are required for indirect estimation: activity-based water requirements and water use coefficients, correlation-based models estimated with regression techniques, econometric models of water use behavior, materials flow models estimated from macroeconomic national accounts data, system-level models that explicitly optimize water use decision-making, and indirect methods for consumptive use.

Coefficient-Based Methods

Coefficient methods or unit use coefficient methods (including the well-known per capita methods) are widely utilized and are particularly well established in urban and municipal water use planning (Billings and Jones, 1996). Commonly used coefficient methods estimate water use as a constant requirement that scales linearly with a physical unit of activity, a dollar value of output, or the size of the water-using population. A more robust coefficient-based model calculates water use separately for specific categories (e.g., commercial, residential, industrial). When adequate categorical data are available, these estimates can be reaggregated to yield more accurate estimates of total water use.

Coefficient-based methods assume constant water use rates in each category of use. This simplification ignores trends, changes in water use due to conservation, technological change, or economic forces, and the optimal level of disaggregation of water use categories. The accuracy of coefficient-based estimates depends both on the water use coefficient and on the underlying activity assumed to drive water use (see Box 4.2). For example, in estimating future water use for the Washington, D.C., metropolitan area, Mullusky et al. (1995) used water use coefficients for three categories of water users: single-family homes, multiple-family homes, and employment water use. These categories were adopted not because of the a priori quality and accuracy of the water use coefficients, but because the underlying housing and employment data were uniformly and consistently available with high spatial resolution (Woodwell and Desjardin, 1995).

BOX 4.2
Error Analysis of Coefficient Methods

Coefficient methods estimate water use, W, as the product of a relevant explanatory variable X (number of employees, number of single-family homes, etc.) and a dimensionally consistent water use coefficient C (gallons per employee, gallons per single family home): $W = XC$. Coefficient or per capita water use estimation implicitly assumes all other explanatory variables either are irrelevant or are perfectly correlated with the single driving variable.

A more robust estimation framework incorporates both X and C as random variables, with means μ_C and μ_X, standard deviations σ_C and σ_X, and correlation ρ_{XC}, making the water use estimate a function of these random variables whose properties can be explored. Considering the covariance of X and C

$$Cov(XC) = E\left[(X - \mu_x)(C - \mu_c)\right] = E[XC] - E[X]E[C] \qquad (4.1)$$

where the expected value of water use $E[W]$ is given by

$$E[W] = E[XC] = \mu_x \mu_c + \rho \sigma_x \sigma_c \qquad (4.2)$$

from which it is clear that the naive water use estimator $E[W] = \mu_X \mu_C$ is biased unless X and C are uncorrelated.

This simple treatment of uncertainty expands water use science in at least two ways. First, the explicit treatment of randomness in these planning-level estimators identifies a source of bias in the estimate and gives an analytical expression for that bias. Second, this simple analysis shows the importance of correlation between X and C, framing substantive questions for water science research and testable hypotheses that can be evaluated through well designed targeted data collection.

For example, Equation 4.2 raises the significant question, "Is a nonzero correlation between X and C plausible?" Consider the simple coefficient estimate of employee water use in which X is the number of employees and C is the per-employee water use. Employee water use is commonly estimated with a water use coefficient, in part because employee data in one form or another are often readily available. It is easy to imagine C varying from establishment to establishment, so treating the water use coefficient as a random variable is fairly intuitive. The randomness in employment (or number of single-family homes, acres, etc.)

The simple example in Box 4.2 shows how analyzing the properties and error characteristics of water use estimators can help structure hypothesis-driven water use science that can be tested in USGS district water use programs. This science improves our understanding of water use, with immediate returns in improved techniques for water use estimation and uncertainty analysis.

may be less intuitive because data such as data on employment often come from published sources as a single value. Nevertheless, recognizing the interpretation of X as a mean level of employment for the estimation period, any single value can only be an estimate of the constantly fluctuating employment base.

Is it plausible for these two random variables to be correlated? Consider employment in a high-density office building. One can easily imagine a component of water use that is insensitive to employee numbers, such as water used for landscaping, evaporative cooling, cleaning, and building maintenance. Combining a relatively invariant water use with individual employee use varying linearly with employee numbers, one might reasonably postulate a negative correlation between X and C. This negative correlation immediately suggests a testable hypothesis that can be evaluated with a sound experimental design for data collection.

Such applied water use science could be effectively coordinated among district water use programs throughout the country to stratify employee water use data by commercial density, climate zone, domestic product, etc. Indeed, the hypothesis-driven nature of the applied research strongly suggests the type of explanatory variables for which data should be collected in this experiment. The water science results would be expected to resolve the question of correlation between X and C, and to clarify some of the explanatory factors and climatic differences in these correlation patterns. These results would also feed directly back into the rule base for water use estimation as modified rules that now use Equation 4.2 as the improved water use estimator with the correlation, drawn from a database developed through targeted data collection, driven by hypothesis-based water use science.

Finally, having resolved this meaningful question of correlation, confidence limits on the coefficient-based estimate can also be computed. Schwartz and Naiman (1999) showed that a distribution free estimate of the standard deviation of equation 4.2 is given by:

$$\sigma_w = \sqrt{\sigma_c^2 \mu_x^2 + \sigma_x^2 \mu_c^2 + 2\rho_{cx}\sigma_c\sigma_x\mu_c\mu_x} \qquad (4.3)$$

With the stronger assumption that the joint distribution of X and C are lognormal, the standard deviation of the water use estimate is given by:

$$\sigma_w = \sqrt{\mu_c^2\mu_x^2\left(1+\rho_{cx}v_cv_x\right)^2\left[\left(1+v_c^2\right)\left(1+v_x^2\right)\left(1+\rho_{cx}v_cv_x\right)^2 - 1\right]} \qquad (4.4)$$

where v_x and v_c are the coefficients of variation of X and C, respectively.

Multivariate Regression

Multivariate regression methods, such as those described in Chapter 6, systematically utilize the correlation structure of observed water use and explanatory variables under the familiar assumptions of the standard linear model. For example, the regression of crop yields on evapotranspiration is widely used in irrigation planning (Nielsen, 1995). In general, multivariate regression models are variations of "water requirement" models in which water use is computed at a fixed, activity-specific rate that is estimated statistically from observed data. For example, most of the observed variability in municipal water use can be explained by exogenous variables representing population and meteorology. Ancillary explanatory variables such as disaggregated data on housing price, per capita income, lot size, and water rates can further explain the variance in observed water use, when such data are available. Of course, explanatory variables such as property value do not directly *cause* water use. Rather, explanatory variables serve as readily available covariates of water-using activities that tend to be strongly correlated with the readily available data such as data on property value.

The most significant issues in modeling water use with multivariate regression are typically (1) selection of the "best" set of explanatory variables, (2) the form of the regression equation, and (3) parameter estimation. The choice of explanatory variables artfully balances plausible causal factors with the observed correlation structure of available data. The well-known challenges of distinguishing correlation and causality particularly apply to water use. Price sensitivity, technological innovation, and scarcity can result in significant changes over time that may easily be misinterpreted as a trend because of their gradual effect and consequent correlation with time. For example, the dramatic increases in manufacturing and industrial water use efficiency over the last 30 years are well documented (see Table 4.1). This "trend" is largely explained as the response to wastewater treatment costs associated with passage of the Clean Water Act in 1972. This historical change in industrial water use would be correlated with a variable representing time. However, a regression model using time as an explanatory variable unrealistically assumes that the dramatic reductions in water use that have resulted from technological process changes and recycling will continue in the future at the same rate.

Similarly, the form of the regression equation should reflect the structural relationship between water use and explanatory variables—not just their linear

TABLE 4.1 Recycling Rates for Various Industries in the U.S.

Industry	1954	2000
Paper	2.4	11.8
Chemicals	1.6	28.0
Petroleum	3.3	32.7
Primary Metals	1.3	12.3
Manufacturing	1.8	17.1

SOURCE: Thompson (1999).

correlation. Municipal water use typically correlates to seasonal changes in temperature and precipitation. Hot, dry weather tends to increase water use. However, in comparing municipal water use in Northern Virginia and Los Angeles, Whitcomb (1988) found that the climatic differences between the two systems required fundamentally different water use models. In temperate Northern Virginia, it is uncommon to have more than five consecutive summer days without measurable precipitation. The most parsimonious model for Northern Virginia water use explained increases above the mean seasonal water use by the number of consecutive antecedent dry days. In contrast, during the dry season in Los Angeles, mean seasonal water use is insensitive to the number of antecedent dry days, but unusually cool or wet weather can decrease municipal water use from the seasonal mean. Although meteorological anomalies are generally correlated with water use, model specification for multivariate and time series models must be carefully matched to system characteristics.

Finally, the complex correlation structure of water use and candidate explanatory variables almost always violates the assumptions of the standard linear model, motivating substantial research on alternatives to ordinary least squares for parameter estimation. This becomes necessary when the most highly correlated explanatory variables are themselves both correlated to, and functions of, water use. For example, the Massachusetts Water Authority (MWA) undertook an aggressive conservation program when water use grew to exceed the yield of Quabbin Reservoir. Water use was successfully reduced through a combination of industrial process changes in the metal plating industry, correction of a leak detected in a distribution system with vintage water mains, and residential water audits. The intensity of MWA's water conservation program (quantifiable by measures such as the annual program budget or the number of water-conserving devices installed) will be correlated with reductions in water use. The intensity of the conservation program will also be a *function* of water use—with greater water use eliciting greater investment in conservation. The effect of the residential conservation program is further confounded by the fact that water rates dramatically increased at the same time (to finance massive infrastructure investment in wastewater treatment) and are therefore correlated with conservation investment. This complexity, typical in water use estimation, suggests the need to model water use as more than just a material requirement correlated with activity indicators. Water-using behavior is the result of consumer choices and policy interventions that can be understood using econometric methods.

Econometric Methods

Econometric methods also estimate water demand using regression-based techniques. However, an econometric approach to water use estimation implies more than including economic explanatory variables in multivariate regression. An econometric approach to water use estimation starts from the fundamental premise that the *demand function* for water is the solution to consumers' constrained optimization problems and therefore varies with the price both of water and all other goods and services. The economic demand function represents water-consuming *behavior*. In contrast, water requirement or empirical regression models implicitly treat water as a fixed technical requirement.

The contrast between regression and econometric approaches to water use estimation can be highlighted by considering the choice of explanatory variables used to estimate municipal water use. A typical model of municipal water use might include variables representing population, weather, lot size, household income, and binary "dummy variables" indicating the presence or absence of swimming pools and water-conserving fixtures (Billings and Jones, 1996). A variable such as the price of electricity would be unlikely to appear in such a model. In contrast, the cost of electricity might be a natural variable to include in an econometric model of water use, because water and electricity are both critical inputs to household activities, and their relative prices influence consumption decisions. For example, the peak supply problems affecting the electrical grid in California in 2001 have led to calls for consumers to conserve electricity by only running dishwashers and washing machines with full loads. Such a behavioral change could be expected to result in a marginal decrease in water use. This decrease would be expected not because of collateral stimulation of consumers' conservation ethic, but rather because water and electricity are *complementary inputs* to the production of dishwashing and laundry services. Their joint consumption is defined by the "production function" for these activities.

Water and electricity may also be *substitutes* in providing cooling and air conditioning services. Residential evaporative cooling—commonly used, for example, in the city of Phoenix—requires only a third to a fifth the electricity of air conditioning. This reduced power consumption is associated with substantial—and highly variable—water use requirements, variously estimated between 95 and 200 gallons per day (Karpisack and Marion, 1994). Indeed, the adoption of evaporative cooling (with its associated increase in water use) was promoted as a means of reducing electricity consumption following escalating energy costs in the 1970s. Water is a direct substitute for energy in this application. However, Karpisack and Marion (1994) point out that, indirectly, water and electricity are complementary inputs, because water is used to produce electricity, and electricity is required to treat, pump, and deliver water.

The economic demand function for water may also be derived as the solution to an optimization problem that explicitly incorporates the underlying water-consuming technological choices and economic objectives driving water use decisions (Gisser, 1970). Such explicit optimization approaches have been extensively used in analyzing water use in California's Central Valley. For example, Howitt et al. (1999) used the Statewide Water Agricultural Production model (SWAP) to quantify the potential gains from exploiting the spatial and temporal differences in the marginal valuation of water. They did this by assessing agricultural users' willingness to pay for reliable water supply. For 25 regional production areas (including the 21 regions of the Central Valley Production Model plus southern California), water application is determined from the solution of regional constrained optimization problems that allocate land, water, and capital to equate the marginal product and marginal cost of water use (Howitt et al., 1999). The total amount of irrigated land in production varies with both the price and availability of water, reflecting the marginal expansion and contraction of irrigated acres as junior water rights are fulfilled.

National Accounts And Input-Output Modeling

The role of water as a commodity and economic input can be quantified through the system of national industrial accounts maintained by the Bureau of Economic Analysis (BEA). BEA consistently quantifies economic flows appearing as market transactions in the national economy at the county, regional, and national levels (Figure 4.1).

Economic activity is quantified as the dollar value of goods and services supplied to industries and final consumers, disaggregated by Standard Industrial Classification (SIC) Codes—or by the North American Industry Classification System (NAICS). One potential use of this uniformly available data is the estimation of water use through the direct use of economic flows reported from the different industrial uses corresponding to the water industry. SIC codes 4941 and 4971 include water supply and irrigation services, respectively. Clearly, the dollar value of market transactions from these industrial sectors reflects total economic output, not just water use. Moreover, significant nonmarket water uses (such as self-supplied wells and instream flows) are not well documented in these national reports. Nevertheless, it may be possible to derive water use coefficients that apply specifically to SIC codes, having units of gallons per dollar. As with all coefficient methods, the quality of the estimate will depend on the quality of both the derived water use coefficient and the estimated magnitude of the underlying activity (i.e., sales). In this case, the consistent availability of economic activity data by sector may warrant the effort to develop appropriate water use coefficients to exploit this high-quality data source.

The BEA also maintains secondary data that track the changes in economic activity throughout the nation. This activity data may be very useful as direct "explanatory variables" for water use estimation. Of perhaps greater value is the potential use of these data to identify significant structural changes and trends in regional economic activity that will require additional scrutiny and investigation to accurately update regional water use estimates. As a consistent indicator of county-level economic activity, BEA's regional economic data may be an efficient starting point in the sampling and design of each cycle of national water use estimation. An example in Table 4.2 shows data on county-level employment, which is uniformly available as part of the BEA Regional Accounts Data. These

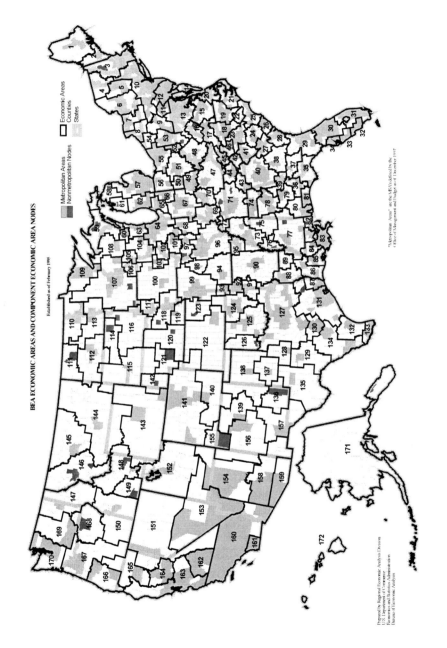

FIGURE 4.1 Bureau of Economic Analysis economic analysis areas.

TABLE 4.2 Full-time and Part-time Employment (Number of Jobs) by Industry, Montgomery County, Maryland, 1994–1998

Industry	Number of Jobs				
	1994	1995	1996	1997	1998
Total	509,120	526,404	532,652	548,047	566,575
Farm	881	876	844	873	784
Non-Farm	508,239	525,528	531,808	547,174	565,791
Private (non-farm)	421,000	436,579	445,955	462,399	480,756
Agriculture	5,480	5,494	5,623	5,678	5,773
Mining	605	646	520	566	558
Construction	28,672	28,823	28,811	29,905	30,197
Manufacturing	17,947	17,700	18,549	20,131	20,428
Transportation and Public Utility	15,449	15,731	14,716	14,845	16,897
Wholesale Trade	15,521	16,122	15,711	15,513	15,417
Retail Trade	76,421	77,764	78,454	80,777	81,640
Finance, Insurance, Real Estate	52,269	55,899	56,776	58,474	60,688
Services	208,636	218,400	226,195	236,510	248,558
Government and Government Enterprises	87,239	88,949	86,453	84,775	85,635
Federal, civilian	44,069	44,682	43,021	41,517	41,289
Military	8,161	8,111	7,834	7,404	7,242
State and Local	35,009	36,156	35,598	35,854	37,104
State	1,673	1,664	1,442	1,202	1,239
Local	33,336	34,492	34,156	34,652	35,865

SOURCE: BEA Regional Accounts Data (http://www.bea.doc.gov/bea/regional/reis/action.cfm).

data would naturally support a simple coefficient-based water use estimation model for "employee water use." The consistent national availability of these county-level data may well justify the effort to refine employee water use coefficient estimates. For example, the pattern and structure of "employee water use" coefficients among the states could be systematically investigated through a series of state-level studies coordinated through the NWUIP, accounting for regional differences in climate and economic trends. The result of these investigations would feed back to the water use program in the form of improved procedures and quantitative uncertainty analysis for national estimation of employee water use with coefficient methods.

INPUT–OUTPUT TABLES AND MATERIALS FLOW ANALYSIS

A disaggregated static "snapshot" of the flows of goods and services in the market economy is also reliably available in national Input-Output (I-O) tables.

I-O tables consist of a matrix showing the interindustry purchases and final sales to users and the commodity inputs required by an industry to produce its output. Detailed I-O tables for the United States are available at several levels of dis-aggregation, with the most detailed accounting for 490 industrial sectors. At the level of so-called "two-digit" industrial codes, the I-O tables aggregate to 97 distinct industrial sectors. Market transactions related to water use are captured in I-O Industry Code 68C: Water and Sanitary Services.

I-O modeling can capture both direct and induced water uses associated with a unit of production. In this way, I-O analysis allows the total water use inputs to be quantified for a unit of output from every industrial type. For example, water is used directly in manufacturing an automobile (e.g., as both process water and for cooling, drinking, and sanitary services within the plant). Automobile pro-duction also uses water indirectly, in the steel, fabric, and other intermediate material inputs to the manufacturing process. Table 4.3 illustrates estimates of embodied water content for various commodities. The I-O table can be used to compute the total (direct and indirect) output required from every industry to produce the disaggregated gross national product of the United States. The I-O tables also represent a set of linear production functions (constraints) for the national economy (under the static assumption of constant technology).

The economic flows quantified in the national I-O tables can also be used to estimate material flows in the economy. Figure 4.2 represents the approach described by Biviano et al. (1999) for estimating the material flows of lead. The methodology draws upon similar efforts by the Department of Defense to esti-mate strategic material requirements and by the U.S. Bureau of Mines (efforts now conducted by the USGS Minerals Information Team). Similar materials flow analyses have been developed for metals and minerals ranging from arsenic cadmium salt and lead to mercury, zinc, and crushed aggregate for concrete. The approach combines National Accounts Data from I-O tables with available mate-rials requirements by industry. The result integrates industry requirements and macroeconomic feedbacks to estimate the total material flow embodied in the final production of each sector of the economy.

TABLE 4.3 Embodied Water Use for Producing Various Products

Product	Water Use (gallons)
Automobile (1)	100,000
Cotton, 1 lb.	2,000
Aluminum, 1 lb.	1,000
Corn, 1 lb.	170
Steel, 1 lb.	25

SOURCE: Thompson (1999).

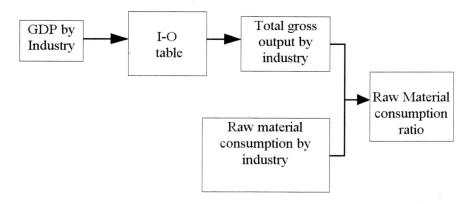

FIGURE 4.2 General methodology for estimating materials flows. GDP is gross domestic product. SOURCE: Biviano et al. (1999).

Although materials flow analysis has primarily focused on minerals, metals, and energy, the combination of BEA economic accounts data and the USGS water use estimates suggests an analogous approach to describing the flow of water through the economy. In the growing field of materials flow analysis and industrial ecology, Wernik and Ausebel (1995) observe that, on a mass basis, the material flow for water is more than 25 times greater than any other material flow in the economy.

Augmented Income and Product Accounts

The existing system of National Income and Product Accounts, along with National Accounts Data and Regional Accounts Data maintained by the BEA, can be an extremely useful source of nationally consistent, county-level data on industry-disaggregated economic activity. The convention that counts only market transactions yields an incomplete representation of national economic activity. For example, cutting both commercially managed and first-growth forests represents drawdowns of national wealth, yet only the former asset loss is recorded in the national accounts. Recognizing the distortions introduced by these conventions, the BEA has begun to develop satellite or "augmented economic accounts" with the goal of accounting for as much economic activity as is feasible, regardless of whether it takes place inside or outside the marketplace (NRC, 1999).

In reviewing the initial and proposed augmentation of the national accounts, the NRC Panel on Integrated Environmental and Economic Accounting found the development of environmental and natural resource accounts to be an essential investment for the nation. The panel concluded that the development of appropriate accounts for the water resources sector should be a high priority. The panel also found that to improve environmental accounting, one must improve the physical measures of stocks and flows on natural resource assets, noting that generally "... there are no routine measures when these flows take place outside of the marketplace" (NRC, 1999). The structure of the USGS water use estimates is particularly well suited to supporting the national accounting for the "near-market" water use services that permeate the economy. Indeed, the USGS water use program was identified as the *only* source of physical data on water use cited by the panel (NRC, 1999). The panel concluded:

> Much of the information needed to construct and maintain environmental accounts would be highly useful to other federal agencies, particularly for natural assets under federal stewardship and for environmental activities for which the federal government has responsibility to undertake benefit-cost analysis. A cooperative and coordinated approach among analytic teams of researchers from different federal agencies and the private sector to collect, analyze, and manage improved natural-resource and environmental data would be valuable not only for developing natural resource and environmental accounts, but also for promoting better monitoring, assessment, and policy making in these areas.

The consistently high-quality data in BEA's national and regional accounts have great potential to serve as a valuable source of consistent data for both screening, trend identification, and water use estimation. As well, the USGS water use program is a unique source of national data on the nonmarket activities derived from the nation's water resources as they flow through both the economy and the hydrologic system. There is great potential for cooperation and coordination between the NWUIP and the BEA, galvanized by the mutual interest in "promoting better monitoring, assessment, and policy making" for the nation.

Indirect Estimates of Consumptive Use

Consumptive use is nearly impossible to measure directly on any scale except the experimental scale. Thus, indirect methods are almost always used.

The simplest method, and a commonly used one, is to estimate consumptive use as the residual of other metered or estimated quantities. For example, consumptive use may be estimated as withdrawals minus return flow for a public-supply system. In this case, consumptive use includes such disparate processes as evaporation from reservoirs or canals, leakage from pipes, public use (street cleaning, fire fighting), theft, etc. A similar calculation could be made for irrigation systems, if conveyance losses can be estimated.

Simple coefficient methods are also commonly used. For example, domestic self-supplied consumptive water use (lawn watering, cooking, etc.) is typically estimated as a percentage of water withdrawn. The withdrawals themselves are typically estimated using per capita coefficients that vary by location. A similar methodology may be used for industrial consumptive use, based on SIC code.

These estimates are highly subject to error depending on the age and specific type of the facility.

Physically based estimates are commonly used for irrigation consumptive use. This is a broad and important topic that cannot be treated in detail here; the reader is referred to Allen et al. (1996) and Jensen et al. (1990) for extensive discussions on the subject. Some of the common methods use empirical correlations between evapotranspiration and climatic factors; others use energy- or water-budget approaches. In the last decade, efforts have been under way to construct spatially integrated fluxes of water by combining eddy flux measurements with satellite measurements and various models. Information on FLUXNET, a global network of such micrometeorological sites, is available at http://www.daac.ornl.gov/FLUXNET/fluxnet.html.

Finally, several of these methods may be combined with remote sensing techniques. For example, The Lower Colorado River Accounting System (LCRAS) estimates and distributes consumptive use by vegetation to water users along a sector of the river, as required by the U.S. Supreme Court decree, 1964, Arizona v. California. LCRAS uses a water-budget method to estimate annual consumptive use by vegetation. Then it combines this output with that from digital-image analysis of satellite images, estimated water use rates by vegetation types, and digitized boundaries for diverters (Owen-Joyce and Raymond, 1996).

CONCLUSIONS AND RECOMMENDATIONS

Compiling a complete inventory of all water use in the United States is intuitively appealing, but impractical. This chapter provides an overview of the diverse methods available to estimate water use when direct measurements and observations are unavailable. All of these methods address both the quantitative estimation of water use and the uncertainty of the estimate. No single approach to water use estimation can be expected to satisfy all the requirements for the NWUIP. Nevertheless, rigorous estimation of water use, matching appropriate methods to available data and objectives, should be a fundamental component of the NWUIP.

The committee therefore recommends that the USGS undertake a systematic national comparison of water use estimation methods in order to identify the techniques that are best matched to the requirements and limitations of the NWUIP. The USGS's structure of state water use specialists and statistical expertise in the National Research Program is ideally suited to this undertaking.

The power of statistical sampling and estimation to rigorously address uncertainty, and to identify and exploit the inherent structure in water use, is illustrated in detail in Chapters 5 and 6.

5

Stratified Random Sampling to Estimate Water Use

Several states with extensive water use databases rely upon the census approach. That is, data are collected for all water users withdrawing amounts greater than a specified threshold that varies from state to state. In some cases—New Jersey, for example—seasonal and annual water use data are collected. In other states, water use quantities may be estimated from empirical equations relating water use to other variables like population and economic activity. In these cases, the water use data may reflect few direct measurements.

The information obtained from the census approach is valuable in understanding water use patterns. However, census data collection may be costly. Indirect estimation techniques allow preparation of aggregated water use estimates, but the quality of the information may be low or uncertain. Therefore, methods that minimize data collection and compilation costs while producing water use estimates with the needed level of accuracy are preferred.

Random sampling is an alternative to exhaustively collecting and processing water use numbers from all users (the census approach) or indirectly estimating use from empirical equations such as coefficient methods. With random sampling, a subset of randomly selected users would complete water use surveys. Statistics derived from the survey results for sampled users would be used to estimate total water use for all users. Compared to the census approach, random sampling reduces the effort involved in collecting water use data, while allowing quantification of the introduced sampling error.

Random sampling is most likely to be useful when done within categories expected to show similarities in the nature of water use. Characteristics of users and geographical location form the basis for dividing users into (mutually exclusive and collectively exhaustive) categories or strata. Within each category, water use data from surveys are averaged. This quantity multiplied by the total number of users in the category produces an estimate of total water use by category. The total water use estimate for a region is just the sum of the total water use in the categories. Theory and techniques of stratified random sampling are discussed in Cochran (1977) and elsewhere. Water use categories are defined in Chapter 3.

The stratified random sampling approach allows explicit estimation of the error due to sampling. Additional error may result from measurement inaccuracies, deliberate misrepresentation of water use, and the failure to identify or appropriately categorize all users. Cochran (1977, Chapter 13) describes some sources of error and discusses needed theoretical modifications.

Stratified random sampling is used successfully by other federal agencies to address similar sampling problems. For example, the U.S. Department of Agriculture (USDA) uses a stratified random sampling approach to estimate irrigated acreage nationwide. Irrigators are grouped into strata based on their past reports of acres irrigated. Strata boundaries are flexible and differ from state to state. The USDA publishes its methods and data in the Census of Agriculture (USDA, 1999a).

STRATIFIED RANDOM SAMPLING METHODOLOGY

The following discussion summarizes some notation and relationships used to estimate water use with stratified random sampling without providing details and derivations. Readers interested in a full analysis of the equations are referred to Cochran (1977).

The fundamental goal of the stratified random sampling methodology is to develop an estimate of total water use, $\hat{Y}$, using data collected within categories. Let the index h represent the hth stratum/category in a water use survey, where h = 1, 2, 3, . . . ,L and L is the total number of strata. N_h is the number of users in category h, and $\sum_{h=1}^{L} N_h$ is the total number of users, N. Let n_h be the number of samples taken from stratum h, and further, assume that water use sampled from a single water use site within each category has population variance σ_h^2.

Then from Cochran (1977, Equation 5.6) the error variance V_T (the square of the standard error for the total water use) is:[1]

$$V_T = \sum_{h=1}^{L} \frac{N_h^2 \sigma_h^2}{n_h} - \sum_{h=1}^{L} N_h \sigma_h^2 \qquad 5.1$$

---

[1]The variance in the mean has been written in terms of V_T since $V_T = N^2 V(\bar{y}_{st})$; $\bar{y}_{st}$ is the estimate for the average water use as used by Cochran (1977).

Note that this equation differs from equations commonly used to estimate variance for samples drawn from infinite populations. The second term on the right side of Equation 5.1 represents the finite population correction, which is essential in this situation because populations of water users are finite. V_T is the error variance attributable to sampling—i.e., the variance that can be controlled by increasing or decreasing sample size. Thus, if every user is sampled, $n_h = N_h$ and V_T equals zero.

The optimal allocation of samples between strata depends upon the total number of users in each category and the population variance within each category.[2] For stratum h, the optimal number of samples, n_h, can be calculated with the following (notation modified from Cochran, 1977, Equation 5.26):[3]

$$n_h = n \frac{N_h \sigma_h}{\sum\limits_{h=1}^{L} N_h \sigma_h} \text{ for } h = 1, 2, 3, \ldots, L \qquad 5.2$$

where n is the total number of samples needed to estimate water use to the desired precision. The total n depends on N_h, σ_h^2, and V_T, the error variance in the total $\hat{Y}$:[4]

$$n = \frac{\left[\sum\limits_{h=1}^{L} N_h \sigma_h\right]^2}{V_T + \sum\limits_{h=1}^{L} N_h \sigma_h^2} \qquad 5.3$$

Equation 5.3 is only valid if samples are allocated in accordance with equation 5.2.

In contrast, if random sampling is performed without stratification, the required number of samples needed is:[5]

$$n = \frac{N^2 \sigma^2}{V_T + N \sigma^2} \qquad 5.4$$

---

[2]Assuming that the cost of sampling does not vary from category to category. Cochran (1977) provides a modification if sampling costs *do* depend on category.

[3]In actuality, Cochran (1977) developed the result in Equation 5.2 by minimizing variance on the estimate of the mean. The result is identical if calculations are made to minimize variance on a total like total water use.

[4]This equation is modified from Cochran (1977, Equation 5.25), where V_T has been used in place of V, the variance in the mean since $V_T = N^2 V$.

[5]This equation is obtained by solving Cochran (1977, Equation 2.8) for n and using V_T rather than V.

where σ^2 is the population variance of samples taken from individual water use sites.

EXAMPLE: DEVELOPMENT OF A SAMPLING PLAN FOR ARKANSAS

The purpose of this example is to illustrate how a sampling approach can be used to estimate total annual withdrawals of water in the state of Arkansas. The example utilizes the existing inventory of point withdrawals within the state, which in 1997 contained 44,670 individual withdrawal points.

The 1997 Database

The 1997 database contains monthly and annual values for 44,670 ground-water and surface water withdrawal points. The information used in the analysis below includes county name, annual withdrawal, source designation, category of use, and pipe diameter. Table 5.1 contains the summary statistics for the 1997 data.

Irrigation withdrawals accounted for nearly 71 percent of the total offstream withdrawals and 92 percent of the total withdrawal points in 1997. A single withdrawal point for nuclear power accounted for nearly 12 percent of the withdrawn volume of water.

Sampling Approach

The data in Table 5.1 were obtained by using a census approach. Although state law mandates the inventory, the inventory entails a cost borne by individual water users and the state government. For all practical purposes, the 1997 data can be taken to represent the entire population of individual withdrawal points in Arkansas, providing us with accurate knowledge of population variances needed to develop a sampling plan. In usual practice, the population standard deviations σ_h would be unknown and would be estimated with sample standard deviations or historical population standard deviations.

Assume that the allowable standard error $\sqrt{V_T}$ due to random sampling is approximately 10 percent of the total annual withdrawal from all categories (12,688,688 MG) or 1,268,868.8 MG. If the total withdrawal were estimated by taking a random sample from the population of all points, to measure the total withdrawal with a standard error of 10 percent, the sample size n would have to be (from Equation 5.4):

$$n = \frac{44,670^2(11,861\ \text{MG})^2}{(1,268,868.8\ \text{MG})^2 + 44,670(11,861\ \text{MG})^2} = 35,560$$

TABLE 5.1 Database of Point Withdrawals in Arkansas, 1997

Category of Use	Number of Withdrawal Points	Mean Withdrawal (MG)	Standard Deviation (MG)	Coefficient of Variation	Total Annual Withdrawals (MG)	Points with Zero Use
Irrigation (IR)	41,102	165	492	3.0	6,771,025	5,417
Agriculture (AG)	1,918	211	328	1.6	403,701	193
Water Supply (WS)	1,026	536	3,837	7.2	550,096	455
Industrial (IN)	200	959	3,829	4.0	191,812	41
Commercial (CO)	120	362	1,286	3.6	43,472	82
Fossil Fuel Power (PF)	49	8,520	32,979	3.9	417,487	19
Minerals Extraction (MI)	33	975	5,488	5.6	32,180	18
Nuclear Power (PN)	15	74,869	289,966	3.9	1,123,034	14
Domestic (DO)	4	2.5	5.0	2.0	10.0	3
Waste Treatment (ST)	4	98	113	1.2	390	2
Hydropower (PH)	2	1,560,228	267,112	0.2	3,120,455	0
Unknown	197	178	264	1.5	35,026	44
All Sectors (combined)	44,670	284	11,861	41.8	12,688,688	6,288

NOTE: Total offstream withdrawals (minus instream use for hydropower) in 1997 = 9,568,233 MG.
SOURCE: USGS Arkansas District Office.

The sample size would be 35,560 withdrawal points or almost 80 percent of the population. When the population variance and the number of users are large (the typical case in water use estimation), the standard error for random sampling and the number of samples required are also very large.

However, the error can be reduced by dividing the population into distinct strata (with smaller strata variances). If optimal stratified sampling is employed, using the use categories in Table 5.1 as the strata, the total number of samples n needed to estimate water use with the same standard error is determined as follows:

$$V_T + \sum_{h=1}^{L} N_h \sigma_h^2 =$$

$$(1,268,868.8 \text{ MG})^2 + \left(41,102(492 \text{ MG})^2 + 1,918(328 \text{ MG})^2 + \ldots + 197(264 \text{ MG})^2\right) =$$

$$\left(3.0966 \times 10^{12} \text{ MG}\right)^2$$

and from Equation 5.3,

$$n = \frac{1}{\left(3.0966 \times 10^{12} \text{ MG}\right)^2}\left[41,102(492 \text{ MG}) + 1,918(328 \text{ MG}) + \ldots + 197(264 \text{ MG})\right]^2 = 339.9$$

Thus, stratification has the potential to substantially improve sampling efficiency. Stratified random sampling reduces the number of samples needed by grouping water use quantities likely to be similar. In this case study, for example, large uses by power plants are separated from smaller irrigation uses, removing some of the sampling variance or randomness.

Allocating the samples according to Equation 5.2 results in the required numbers of samples within each category, as shown in the column in Table 5.2 for corrected number of samples. In two categories (hydropower and nuclear power), the calculated number of samples required exceeds the population size. This impossibility is corrected for such cases by requiring that all users be sampled. Adding an additional requirement that each category have a minimum of two samples results in the corrected values for n_h in Table 5.2.

TABLE 5.2 Number of Required Samples, by Category, to Achieve Approximately 10% Standard Error in the Water Use Estimate

Category of Use	Number of Withdrawal Points	Number of Samples Required, n_h			Standard Error (%)
		Calculated	Corrected	Final	
Irrigation (IR)	41,102	211.9	212	330	16
Agriculture (AG)	1,919	6.6	7	10	49
Water Supply (WS)	1,026	41.2	41	64	86
Industrial (IN)	200	8.0	8	12	112
Commercial (CO)	120	1.6	2	3	202
Fossil Fuel Power (PF)	49	16.9	17	26	52
Minerals Extraction (MI)	33	1.9	2	3	310
Nuclear Power (PN)	15	45.6	15	15	0
Domestic (DO)	4	2.10×10^{-4}	2	2	100
Waste Treatment (ST)	4	4.74×10^{-3}	2	2	58
Hydropower (PH)	2	5.6	2	2	0
Unknown	198	0.5	2	2	105
All Sectors	44,670	339.9	312	471	10

Equation 5.1 allows calculation of the standard error for any sample allocation. Using this equation for the corrected n_h shows that the standard error in the estimate of total use is 1,592,912 MG, or about 12.6 percent error. If this value is unacceptable, additional samples could be taken from the 10 categories containing unsampled users. Solving iteratively for the samples required to result in a standard error of 10 percent, the final numbers of required samples (Table 5.2) is obtained. Thus, to achieve 10 percent standard error, the stratified random sampling approach requires 471 samples, less than 1.1 percent of the population. Within each category, random sampling is used to estimate water withdrawals.

The standard error for individual categories is calculated as $V_{T_h} = \dfrac{N_h^2 \sigma_h^2}{n_h} - N_h \sigma_h^2.$

The last column of Table 5.2 shows the standard error by category, as well as the standard error for all sectors combined. Obviously, in the two categories where all users would be sampled, the error is zero (except for unknown measurement errors). Note, however, in all other categories, the standard errors are greater than 10 percent. In some applications, sample planning may also include objectives on allowable errors for individual categories, or allowable errors for total withdrawals by county or region. If so, additional sampling would be required to meet these objectives.

SUBSTRATA DELINEATION

With the stratified random sampling approach, withdrawal measurements are made only on a subset of the population within a category/stratum. The effectiveness of the stratified random sampling approach depends on the variances of the populations within the strata. Where variances within strata are high, it may be useful to divide a stratum into two or more substrata. The equations described early in this chapter could be easily modified to address the situation where boundaries between substrata are obvious and chosen before sampling begins. In other cases, the decision about subdividing a stratum may seem arbitrary, e.g., dividing a range of users into two subcategories of major and minor users. It may be efficient in such a case to sample a smaller proportion of the minor users, particularly when the variability associated with minor users is low. This may occur, for example, when there are large numbers of users with zero withdrawal (see Table 5.1).

Example: Determination of Substrata
Boundaries and Assessment of Errors

The main purpose of this example is to examine the precision of estimating stratum water withdrawals with an alternative sampling approach. This approach combines statistical sampling of minor withdrawal points with a census (i.e., a complete enumeration) of major withdrawal points. It is designed to overcome the problems presented by the very heterogeneous population from which the sample is taken. In contrast to a standard sampling problem in which the sizes of the strata are predefined, this example approach simultaneously selects the sample size and the boundary between the two strata. By using an electronic spreadsheet, it is possible to calculate the population statistics for each stratum while moving the stratum boundary from the largest withdrawal points to the smallest. Points in the top stratum are individually measured (i.e., through a census), and a random sample is taken from all the remaining points in the population. If we assume a target standard error of 10 percent of the total withdrawal for groundwater and for surface water, Equation 5.4 can be used to calculate the required sample size for the bottom stratum.

The approach is applied to a single water use category: the population of Arkansas' 1997 irrigation withdrawals. The 1997 data for Arkansas reported irrigation withdrawal volumes for 41,102 points, of which 5,417 show no water withdrawal in 1997. Table 5.3 summarizes the statistics for the subpopulations of groundwater and surface water withdrawals for irrigation.

Figure 5.1 shows the total sample size (i.e., all points in the top stratum plus sampled points in the bottom stratum) needed to achieve the target standard error as a function of top stratum size for Arkansas' groundwater and surface water irrigation withdrawals. Note the minimum total sample size required for the groundwater subcategory is less than that for surface water subcategory, even though there are over seven times as many groundwater withdrawal points. This occurs because the relative variability is much lower for the groundwater subcategory.

Still, the delineation of a substratum boundary is very effective in reducing the sampling required for surface water. To estimate the total annual surface withdrawal for irrigation with a 10 percent standard error using random sam-

TABLE 5.3 Population Characteristics for Irrigation Withdrawals, Arkansas, 1997

Characteristic	Groundwater	Surface Water
Number of water use sites	36,053	5,049
Total annual volume, MG	5,492,734	1,278,291
Mean annual withdrawal, MG	152	253
Median annual withdrawal, MG	110	90
Standard deviation, MG	162	1,333
Coefficient of variation	1.1	5.3
Number of samples required	110	163
Standard error of estimate, %	10	10

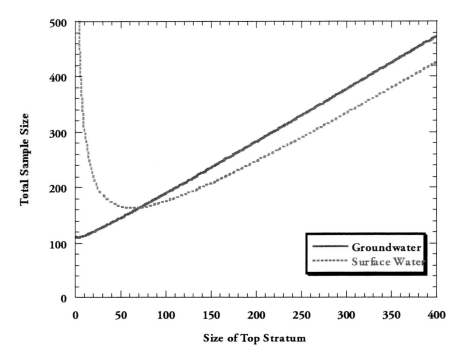

FIGURE 5.1 Optimum sample size for the sampling of groundwater and surface water withdrawals for irrigation using substrata boundaries, Arkansas, 1997.

pling, the required number of samples is 1,789. Using the substratum boundary, the minimum total sample size is 163 (Figure 5.1); the 60 withdrawal points with the largest annual quantities should all be measured (approximately 1.2 percent of all surface withdrawal points for irrigation), and a random sample of 103 withdrawal points (or approximately 2 percent of all points) should be selected from the remaining 4,989 withdrawal points. In contrast, the use of a substratum boundary does not significantly reduce the sampling required for groundwater. Using random sampling, the required number of samples is 110. Using the substratum boundary, the minimum total sample size is 110; the single largest withdrawal points should be measured, and a random sample of 109 points (approximately 0.3 percent) should be selected from the remaining 36,052 groundwater withdrawal points. Hence, with groundwater withdrawals for irrigation, which account for over 80 percent of all withdrawal points in the state database, random sampling is sufficient for statewide water use estimation.

Overall, the division of irrigation withdrawals into two subcategories and the use of substrata boundaries (at least for surface water) greatly improve on random sampling within the irrigation category. As shown in Table 5.2, the use of 330 random samples as part of a statewide sampling plan results in a standard error of 16 percent for the irrigation category. Using the minimum sampling indicated above, a total of 273 samples is required, and with a standard error of about 4.4 percent for the total irrigation withdrawal (using Equation 5.1). Still, it is worth noting that there are some practical problems associated with implementing such a plan. For instance, identifying the largest users could be difficult in the absence of census data. This and other issues would need to be considered as part of the sample planning process.

APPLICATION TO STATES THAT LACK WATER USE DATA

The example analysis used known statistics for 1997 water use in Arkansas to develop an optimal sampling plan that could be used in subsequent years. This example demonstrates the substantial, optimal benefit of stratified random sampling. However, random sampling is most needed for states or programs that currently lack data on water use. In these cases, USGS researchers would not have the information needed to develop an optimal sampling plan. However, stratified random sampling can still be employed as long as it is possible to estimate the number of users in a set of use categories. This is true even in the absence of prior knowledge of the site-specific statistics. The sampling plan developed for this situation would likely be nonoptimal, but stratified random sampling does not have to be optimal to be useful. Using Equation 5.1, it is possible to estimate error variance and standard error for water use estimated from *any* stratified sampling plan. Water use estimates developed from non-optimal sampling plans will be expected to have larger standard errors than estimates developed from optimal sampling plans with the same number of total samples.

The sampling plan itself (values for all n_h) could be established in many different ways when site-specific statistical data are not available. One possibility is to conduct a small preliminary sampling effort. This would allow estimation of category variances needed to prepare an optimal sample plan in accordance with the procedure followed in the Arkansas example. The major difference is that the square root of the preliminary sample variances, s_h^2, would be substituted for the square root of the population variances, σ_h^2, in Equations 5.2 and 5.3. Sample variances s_h^2 derived from the full stratified sample would be substituted in Equation 5.1 to calculate the error variance for the water use estimate. Uncertainty in the sample variances can be estimated using standard statistical procedures and would contribute additional error to the total water use estimate. Uncertainties in the number of users in the categories could similarly be accounted for in each term in the sum. Thus, uncertainty in category size and variance will increase the standard error of the water use estimate. This situation will likely improve as time passes. The statistical data collected during the first complete stratified sampling effort would be used to design an improved sampling plan for future use.

A second possibility for developing a sampling plan in the absence of prior statistics is to use category variances available from state programs with substantive data collection efforts in the optimal sampling plan Equations 5.2 and 5.3 with the target state's category numbers. This will work best when similarities in the categorical water use between the two states are expected. Again, discrepancies between variances in the target state and the data-rich state will render the plan less than optimal, but will not impair the ability to assign an estimated error to the water use estimate.

A third possibility exists where the state has the legal authority to register or permit water withdrawals. Permitted withdrawals could be used as an approximation of actual withdrawals for the purpose of sampling plan design.

This discussion has described only a few ideas for setting up a sampling plan for the first survey when a good water use database is not available. More work is needed to expand this list and evaluate the utility of each option.

The sampling plan would also require information on the number of water users in each category. One approach would be to do a census of the water users. However, this approach might be prohibitive in most states. Alternate methods for estimating the number of water users need to be explored. The Census of Agriculture, the Population Census, manufacturing surveys, and other data sources may provide information necessary for water use population estimation (see Chapter 4). The USGS should also consider consulting with and using the services of experts in other federal or state agencies for help in estimating the number of water users within a state.

ISSUES FOR FURTHER RESEARCH

This is a preliminary study of data from a single state. Comparative studies of data from a number of states with differing degrees of data quality are needed to solidify the understanding of how stratified random sampling can best be applied to water use. Some of the questions that remain to be resolved are the following:

1. Most states have a trigger level for reporting water use, so the population of sites sampled is censored to omit the smallest users. How can stratified random sampling be used to estimate total water use from a sample censored in this manner?

2. A region may have a particular water use category such as irrigation, for which the total number of water withdrawal points is not known precisely and can

only be estimated. How does uncertainty in the size of the total population affect the error of estimate of total water use in the region?

3. What is reported on most water use surveys as the amount withdrawn or used is not a measurement but is itself an estimate by the water user. Can a measure of these site-specific estimation errors be obtained and incorporated within the error estimates for the water use strata so as to adjust the number of users sampled to allow for the estimate of the error inherent in sampling each user?

4. The 50 states, the District of Columbia, and Puerto Rico can be thought of as a collection of 52 sampling units, which each have particular characteristics. An attempt has been made in Chapter 2 to classify these sampling units for data quality. How can the data be examined state by state to more rigorously quantify data quality as a function of characteristics such as the type of laws pertaining to water use data collection?

5. In many states, the samples of water use sites, often obtained by voluntary submission of water use reports, are incomplete. Can incomplete samples of this kind be used in a stratified random sampling framework to arrive at reasonable estimates of the total water use and its standard error?

6. The examples presented in this chapter are all for annual withdrawals in Arkansas. Does the variability of monthly water use differ sufficiently from that of annual water use to require a different sampling plan?

7. Measurement errors for withdrawals were ignored in the Arkansas example. Although stratified sampling reduces the sampling requirements, it still requires quality data for water use estimation. Are modifications needed to the statistical approaches to deal with measurement uncertainty, which might vary from state to state?

The USGS has a strong group of water statisticians who have experience in examining the above issues in other contexts, such as analysis of floods, streamflow, and water quality (Helsel and Hirsch, 1992). The statistical studies recommended here are certainly within their range of competence.

CONCLUSIONS AND RECOMMENDATIONS

This chapter has described some of the potential benefits of using random sampling to estimate water use. Benefits may include reduced sampling and associated costs (compared to performing a full census), simple approaches to assessing the quality of water use estimates, as well as the ability to design a sampling plan to meet particular data-quality needs.

One important benefit of random sampling is a potential reduction in data collection efforts. By incorporating variance reduction techniques, particularly optimal stratified sampling, quality water use estimates may be obtained by sampling less than 5 percent to 10 percent of users. An example analysis for Arkansas showed that only approximately 1 percent of users needed to be sampled for estimating state-level water use, with a standard error of 10 percent. A greater percentage of users would need to be sampled to achieve the same standard error for regional or county-level estimates or for estimates for individual water use categories.

Random sampling facilitates the use of statistical approaches to calculate and report the quality of water use estimates. For example, in this chapter's analysis, standard error in water use estimates was chosen as the measure of quality. Other measures, such as confidence intervals, could also be used to describe quality.

Another advantage of random sampling is that it allows water use program staff to readily design a sampling program to meet particular data-quality needs. In the study performed earlier in this chapter, the total number of samples and the allocation of samples among strata were chosen to meet an allowable standard error. In practice, quality targets could be based on the intended uses for the data, ensuring that sampling efforts and staff resources are directed where they are most needed. Furthermore, sampling programs need not be fixed; dynamic sampling could also be beneficial. For example, if a preliminary analysis of sample data shows that variability in a particular use category is substantially higher than expected, it may make sense to update the number and allocation of samples and perform additional sampling.

Many variations on the random sampling approach are possible. The case study demonstrated the use of a hybrid approach to random sampling within a single category or stratum—samples were drawn randomly for "minor" users in the category, and a census approach was used for "major" users. The study showed that for surface water irrigation withdrawals in Arkansas, the hybrid approach required far fewer surface water samples than with random sampling. However, for groundwater irrigation withdrawals, where the relative variability of withdrawals is much lower than for surface water, the hybrid approach offers no advantage over random sampling. Further guidance is needed to help USGS districts select appropriate statistical sampling techniques for water use estimation in individual states.

The quality of results from the random sampling approach is limited by some of the same issues that affect the census approach. Although full surveys are not required of every use category, it is still critically important to have an accurate count of the total number of users. In addition, when stratified sampling is used, all users must be accurately distributed into categories. Maintaining accurate user counts may require a substantial amount of effort. In addition, whether the census approach or random sampling is used, procedures must be established to manage the response rate (e.g., follow-up for surveys not returned). Depending upon the situation, these efforts could be substantial. As a result, further study is needed to determine whether substantial cost reductions could be achieved through stratified random sampling in actual practice. The examples in this chapter certainly support the possibility of reduced costs and justify an additional investigation.

A disadvantage of reducing the proportion of users sampled (by using random sampling rather than the census approach) is the possibility of increased uncertainty in water use estimates. Thus, it is important to ensure that sampled data are accurate and representative. However, as a result of diminishing returns (each additional sample progressively provides less information as the sample size increases), stratified random sampling has the potential for greatly reducing the data collection workload with small, acceptable increases in uncertainty. This beneficial outcome can be obtained by explicitly balancing costs and accuracy. Reducing the quantity of data collected may even allow increased attention to quality for the fewer data collected.

The committee recommends that the USGS develop statistical sampling approaches for water use estimation as part of the National Water-Use Information Program. Site-specific water use data from various states may be useful in developing and evaluating sampling approaches. The National Handbook of Recommended Methods for Water Data Acquisition (USGS, 2000) and the USGS's internal guides for preparing water use estimates should be updated with a manual of procedures for statistical sampling of water use and determination of total water use estimates and their errors.

6

Regression Models of Water Use

This chapter explores the structure of the past National Water-Use Information Program (NWUIP) state-level aggregated water use data, based on corresponding (and routinely collected) demographic, economic, and climatic data. The purpose of this inquiry is to determine if multiple regression models have the potential to explain the temporal and geographic variability across the United States of the aggregated water use estimates produced by the NWUIP. The statistical models examined here are derived using the U.S. Geological Survey (USGS) estimates of total withdrawals for public supply use and thermoelectric power use. A complete analysis of historical withdrawals is described in Dziegielewski (2002a).

NATIONAL WATER USE DATA

Total water use in the United States has been estimated by the USGS every five years since 1950. National estimates focus primarily on measuring total water withdrawals, which include the annual extractions of both fresh water (with separate estimates for surface water and groundwater withdrawals) and saline water. The total withdrawals are subdivided into categories; all point withdrawals are aggregated and reported at the county and state levels. The structure of these reported withdrawals in 1995 (Solley et al., 1998) can be represented as:

$$TW_t = \sum_i \left(PS_{it} + DM_{it} + CM_{it} + IR_{it} + LS_{it} + IN_{it} + MN_{it} + TE_{it} \right) \quad (6.1)$$

where

$TW_t =$ total (fresh and saline) water withdrawals in all states, the District of Columbia, Puerto Rico, and the U.S. Virgin Islands in million gallons per day (MGD) during calendar year t

$PS_{it} =$ public supply withdrawals (in state i during year t) , MGD

$DM_{it} =$ domestic (self-supplied) withdrawals, MGD

$CM_{it} =$ commercial (self-supplied) withdrawals, MGD

$IR_{it} =$ irrigation withdrawals, MGD

$LS_{it} =$ livestock withdrawals, MGD

$IN_{it} =$ industrial (self-supplied) withdrawals, MGD

$MN_{it} =$ mining withdrawals, MGD

$TE_{it} =$ thermoelectric withdrawals, MGD

In the 1995 compilation, freshwater withdrawals were estimated for all eight categories (or sectors), and saline water withdrawals were estimated for industrial, mining, and thermoelectric categories. The freshwater withdrawals are separated into groundwater and surface water for all sectors, and saline withdrawals are separated by source for industrial, mining, and thermoelectric sectors. For example, the total withdrawals for thermoelectric power use, TE_t , can be represented as:

$$TE_t = \sum_i \left(TE_{itfs} + TE_{itfg} + TE_{itbs} + TE_{itbg} \right) \qquad (6.2)$$

where

$TE_{it} =$ withdrawal for thermoelectric power use in state i during year t; and the subscripts f, b, s, and g respectively indicate freshwater, brackish or saline water, surface water, and groundwater.

These eight categories are nonoverlapping and sum up to total withdrawals. However, public supply withdrawals include water delivered by public water supply systems to some commercial, industrial, and thermoelectric uses, and detailed sectoral-use tables in Solley et al. (1998) show both the self-supplied withdrawals and deliveries of water to each sector.

The reported estimates are obtained primarily from detailed inventories of point withdrawals within each accounting unit (i.e., county or state). The point withdrawals represent measured volumes of water at pumping or diversion points or estimates of the withdrawn volumes based on the time of pump operation, irrigated acreage, or some other indirect measure. Indirect measures depend on water use category and assume a specific relationship between the quantities of water use and the values of the corresponding indirect measures (USGS, 2000, Chapter 11). Statistical models of water use permit an explicit consideration of the relationships between water use and these indirect measures. These relationships are discussed in the following section.

WATER USE RELATIONSHIPS

Water use at the state level can be estimated indirectly by using multiple regression analysis. In regression models, water use relationships are expressed in the form of mathematical equations, showing water use as a mathematical function of one or more independent (explanatory) variables. The mathematical form (e.g., linear, multiplicative, exponential) and the selection of the right-hand-side (RHS) or independent variables depend on the category and on aggregation of water demand represented by the left-hand-side (LHS) or dependent variable. A large number of econometric studies of water use have been conducted. Hanemann (1998) summarizes the theoretical underpinnings of water demand modeling and reviews a number of determinants of water demand in major economic sectors. Useful summaries of econometric studies of water demand can be found in Boland et al. (1984). Dziegielewski et al. (2002b) reviewed a number of studies of aggregated sectoral and regional demand. A substantial body of work on model structure and estimation methods was performed by the USGS (Helsel and Hirsch, 1992).

Depending on the purpose for which the estimates are used, the dependent variable (i.e., water use) can be presented in different ways. For example, in studies of surface and groundwater resources, the data are usually available as daily, monthly, or yearly withdrawals at a point such as a river intake or a well. Because the water withdrawn is typically used (or applied) over a larger land area, an equivalent hydrologic definition of water use would be the use of water over a defined geographical area (e.g., an urban area, a county, or a river basin). As shown in Equation 6.1, total water use within a larger geographical area such as a county or state can be presented as a sum of water use by several groups of users within a number of subareas.

Generally, water use at any level of aggregation can be modeled as a function of one or more explanatory variables. However, the best results are obtained by breaking down total water use by sector, because different subsets or explanatory variables apply to different sectors. For example, public supply withdrawals can be estimated using the following linear model:

$$PS_{it} = a + \sum_j b_j X_{itj} + \varepsilon_{it} \tag{6.3}$$

where PS_{it} represents public supply withdrawals within geographical area i during year t, X_j is a set of j explanatory variables (e.g., air temperature, precipitation, price of water, median household income, and others), which are expected to explain public supply withdrawals, and ε_{it} is a random error term. The coefficients a and b_j can be estimated by fitting a multiple regression model to the historical data. This procedure has some parallels in modeling river loads, sediment-rating curves, and urban nonpoint pollution loads. Examples of studies of those subjects, which utilize statistical approaches, include Cohn et al. (1989) and Christensen et al. (2000).

WEATHER NORMALIZATION OF WATER USE

The quantity of water withdrawn in any given year depends on weather conditions. Water withdrawals for most purposes increase during periods of hot and dry weather and decrease during periods of cool and wet weather. This dependence of withdrawals on weather conditions can be determined by including weather-related variables in the set of explanatory variables X_j in Equation 6.3 above.

The accuracy of the weather adjustment depends on the length of the time interval used in data averaging. The best results are obtained by modeling time-series data on daily or weekly water use; the relationship can be masked when monthly and seasonal data are used. For example, water use is negatively correlated with precipitation. However, if monthly data are used, it is possible that total precipitation during a given month could be higher than normal but concentrated during the last two days of the month. Water use during that month would be higher than normal because of the dry conditions during all but the last two days of the month, thus indicating a misleading positive correlation between water use and precipitation.

The selection of variables to represent weather conditions depends on the sector. In models of domestic demands, commonly used measures of weather conditions include antecedent precipitation (or antecedent rainless days) and air temperature. Evapotranspiration is often used in models of water use for landscape watering and irrigation demands, and cooling degree-days and heating degree-days are used to estimate industrial demands or thermoelectric power use (Boland et al., 1984; Dziegielewski et al., 1996).

The use of weather variables in multiple regression models is illustrated in the later sections of this chapter. The next section explores the structure of water demand in public supply sector water use and presents several statistical models that were fitted to the historical estimates of public supply withdrawals in the lower 48 states.

STATE-LEVEL MODELS OF PUBLIC SUPPLY WITHDRAWALS

Public supply water is water withdrawn by public or private water suppliers and delivered to users. The public supply withdrawals estimated by the NWUIP for the years 1980, 1985, 1990 and 1995 in each of the lower 48 states were used in regression analysis. Twenty-one variables were selected as the likely predictors of public supply withdrawals at the state level and the following:

• *Population:* resident state population, population served, population density, and percent urban population;
• *Income:* median family income, state per capita income;
• *Economy/employment:* civilian labor force, gross state product per capita, average (weighted) price of water;
• *Housing mix:* percentages of single-family housing units, multifamily housing units, and mobile homes;
• *Weather:* total precipitation (during growing season), average air temperature (during growing season), and extreme monthly value of Palmer Drought Severity Index (PDSI); and
• *State water law:* prior appropriation, riparian or riparian with permits.

These variables are measures of demographics, affluence, economic activity, housing stock, weather, and water allocation arrangements. Six indicator (binary) variables were constructed to represent the legal systems of water rights in each state for allocating surface water and groundwater to uses and users. A measure of "dryness" for weather conditions was chosen as the lowest monthly value of the PDSI during the data year for each state. PDSI may have significant limitations in capturing the effects of dry weather on water use and has been found not to be a nationally consistent measure of dryness (Alley, 1984; Guttman et al., 1992). There are other indicators of the evaporative demand of the atmosphere as it affects the consumptive use of water (e.g., Class A pan evaporation, reference crop evapotranspiration); however, the availability of such measures at the geographical scales used in this analysis is limited.

Population served by public water supply systems was used to express the dependent variable as average public supply withdrawal per capita per day for each state and data year. If the per capita rate of withdrawal in each state can be predicted with sufficient accuracy, then total public supply withdrawals can be estimated by multiplying the per capita withdrawal by population served.

One advantage of modeling the per capita withdrawal is that by expressing total withdrawals in per capita terms, the dependent variable is "normalized" across states, and the problems associated with heterogeneity of total withdrawals among the states are avoided. Also, the "out of range" values of per capita withdrawal can be easily spotted in the data and investigated. It should be noted, however, that regression analysis can also be applied to total public supply, not just to per capita public supply withdrawals as described here.

It should be emphasized that the regression models presented here are for illustrative purposes only, as many details about model diagnostics and other aspects of the analysis have been omitted for clarity. Detailed discussions about potential bias in the estimators and alternative estimation techniques are described in Dziegielewski et al. (2002a).

Table 6.1 shows the coefficients of a linear regression (see Equation 6.3) of 1980–1995 state-level data (excluding the District of Columbia) on per capita public supply withdrawals using the ordinary least squares (OLS) procedure. The shorter data series for 1980–1995 was selected to take advantage of improved data collection procedures and to capture the recent trend of declining water use since the 1980 compilation.

The model shown in Table 6.1 explained 52 percent of the variance in per capita usage rates among states and across reporting years. The predictive properties (regression fit) of the model are limited as indicated by both the absolute and relative size of the residuals shown below the table. The mean absolute percentage error (APE) is 12.9 percent, and the root mean squared error is 31.6 gallons per capita per day (gpcd).

Despite the significant unexplained variance, the regression model in Table 6.1 can be considered to be a reasonable "explanatory" model, which reveals the structure of demand for public water supply even in the geographically aggregated data. The size and signs of the estimated regression coefficients fall within the ranges of expected values. These coefficients can be interpreted to mean that across the United States, from 1980 to 1995, the mean withdrawal was 183.7

TABLE 6.1 Linear Regression Model for State-Level Per-Capita Public Supply Withdrawals, 1980–1995

Dependent/Explanatory Variable	Regression Coefficient	*t*-Ratio	*F*-value Probability
Intercept (gpcd)	115.881	3.28	0.0012
Average price of water			
($/1,000 gal., real 1995 dollars)	–7.779	–2.63	0.0091
Gross State Product per capita			
($1,000, real 1995 dollars)	1.676	3.22	0.0015
Precipitation in summer months			
(May to Sept., in inches)	–2.119	–4.02	0.0001
Average temperature during summer			
(Fahrenheit degrees)	0.983	2.15	0.0326
Indicator of states with prior appropriation			
groundwater rights system	29.136	3.05	0.0027
Indicator of states with prior appropriation			
surface water rights system	17.218	1.81	0.0716

NOTES: Mean water use = 183.7 gpcd; $n = 192$; $R^2 = 0.52$; mean APE = 12.9%; root MSE = 31.6 gpcd; Nine observations of per capita withdrawal in the original data were adjusted using a data-smoothing procedure.

gpcd (from the data). This average withdrawal rate would decrease by 7.8 gpcd if price were increased by $1/1,000 gallons, and it would increase by 1.7 gpcd if the gross state product per capita increased by $1,000. Because a significant portion of public supply withdrawals is used to supply industrial and commercial uses, the gross state product variable captures the effects of the relative volume of nonresidential uses together with the effect of the ability to pay for water, which is typically captured by per capita or median household income variables in models of residential use. The binary indicator variable, which assumes the value of 1 for states with prior appropriation groundwater rights (generally western states), indicates that on average, these states withdrew 29 gpcd more than states with riparian and riparian with permits systems. Also, in states with prior appropriation surface water rights, average per capita withdrawals were on average higher by 17.2 gpcd than in riparian law states. The water rights variables most likely are an indirect measure of the arid climate of the states that use the prior appropriation system rather than indicating increased use because of appropriation rights.

The effects of individual explanatory variables can be also expressed in terms of elasticity of water demand with respect to changes in the values of each dependent variable. Elasticity measures the percentage of change in the independent variable that would be caused by a 1.0 percent increase in the value of independent variable. For example, the elasticity of demand with respect to price (estimated at the means) is –0.10. This value is found by multiplying the regression coefficient –7.779 by the ratio of average price to average per capita withdrawal in the data. An elasticity of –0.10 is relatively low (in absolute value), but it is close to expectation for aggregate public supply data. Also, the elasticity of demand with respect to income (as represented by gross state product) is +0.22. These elasticity values indicate that a 1.0 percent increase in price would result in a 0.10 percent decrease in demand while a 1.0 percent increase in per capita gross state product would result in a 0.22 percent increase in demand.

The estimated regression coefficients for temperature and precipitation in Table 6.1 clearly show the effect of weather on withdrawals and can be used in normalizing water use for weather. In this context, withdrawals during normal weather could be predicted by substituting into the regression equation "normal" values of average air temperature during summer months and total precipitation during the growing season for these dependent variables. The regression coefficients of the two weather variables in the model indicate that the average per capita demand in a state decreases by 2.1 gallons per day (gpd) per one-inch increase in precipitation during the growing season (elasticity at the mean is –0.19). The per capita demand increases by approximately 1 gpd per one-degree increase in average annual temperature (elasticity at the mean is +0.37). These elasticity values indicate that per capita public supply withdrawals decrease by 0.19 percent for each one percent increase in precipitation and increase by 0.37 percent for each one percent increase in average temperature.

The predictions from the model in Table 6.1 can be improved by supplementing them with information that is contained in model residuals (i.e., differences between actual and predicted values). This can be done by introducing binary variables, which designate individual states. In a model with binary state indicator variables, the average value of residuals for each state is added to the predicted value for that state thus reducing the prediction error. Similarly, if the state residuals contain an increasing or decreasing time trend, such a state-specific trend can also be added to the prediction. However, the addition of separate intercepts and time trends for some states does increase the number of model parameters. If the resulting model is overspecified, the coefficients of the continuous variables, which form the structural component of the model, may be biased. Such bias is small when the inclusion of a state-specific intercept (or trend) does not result in an appreciable change in the value of the estimated coefficients of the structural variables. Still, as with any statistical model, careful evaluation of the model predictions is recommended before accepting the final form of model.

An alternative model was fitted using a stepwise procedure that selected the best explanatory variables from both the continuous variables used in the model shown in Table 6.1 and the binary variables, which designate individual states. In addition, a time trend variable was fitted to the data with trend adjustments for several individual states. The model was estimated using a truncated subset of data for 1980, 1985, and 1990, which excluded the 1995 data. The estimated regression coefficients and other related information for this extended model are shown in Table 6.2.

An estimate of per capita public supply withdrawals for any state and year can be made using the model in Table 6.2. This can be done by substituting the corresponding values of price, per capita gross state product, total summer precipitation, and average temperature and adding four "intercept adjustors"—one for state groundwater law system, one for state surface water law system, one indicator of an individual state (if present in the model), and one state-specific trend (if present)—using the following equation:

$$PS_{it} = 90.659 - 4.726AP_{it} + 2.430GP_{it} - 1.299R_{it} + 0.777T_{it} \qquad (6.4)$$
$$+ 17.356LG_{it} + 38.697LS_{it} + a_iS_i + b_iT_iD_i$$

where

$PS_{it} =$ per capita withdrawal (gallons per day) in state i during year t
$AP_{it} =$ average price in constant 1995 dollars
$GP_{it} =$ gross state product per capita in constant 1995 dollars
$R_{it} =$ total summer season precipitation in inches
$T_{it} =$ average summer temperature, degrees Fahrenheit

TABLE 6.2 "Extended" Per Capita Model of Public Supply Withdrawals, 1980–1990

Variables	Estimate	Std Error	t Ratio	Prob>|t|
Intercept (gpcd)	90.659	23.195	3.91	0.0002
Average price of water ($/1,000 gal)	−4.726	1.624	−2.91	0.0044
Gross state product per capita ($1,000)	2.430	0.352	6.91	<0.0001
Total precipitation during summer, inches	−1.299	0.365	−3.55	0.0006
Average temperature during summer (deg. F)	0.777	0.270	2.88	0.0048
States w/ prior appropr. groundwater rights	17.386	7.529	2.31	0.0229
States w/ prior appropr. surface water rights	38.697	6.980	5.54	<0.0001
Indicator for Alabama	50.543	11.722	4.31	<0.0001
Indicator for California	−47.292	13.000	−3.64	0.0004
Indicator for Connecticut	−29.507	8.065	−3.66	0.0004
Indicator for Delaware	−25.258	7.956	−3.17	0.002
Indicator for Florida	16.950	8.513	1.99	0.0491
Indicator for Idaho	27.171	9.020	3.01	0.0032
Indicator for Kansas	−60.388	9.506	−6.35	<0.0001
Indicator for Massachusetts	−32.888	7.805	−4.21	<0.0001
Indicator for Michigan	16.514	7.894	2.09	0.0389
Indicator for Montana	36.237	8.938	4.05	<0.0001
Indicator for Nevada	80.910	8.395	9.64	<0.0001
Indicator for New Hampshire	−23.742	7.714	−3.08	0.0027
Indicator for New Jersey	−14.228	7.744	−1.84	0.069
Indicator for North Dakota	−104.913	12.410	−8.45	<0.0001
Indicator for Oklahoma	−56.023	12.707	−4.41	<0.0001
Indicator for Oregon	−26.390	8.667	−3.05	0.0029
Indicator for Pennsylvania	33.247	7.521	4.42	<0.0001
Indicator for Rhode Island	−27.130	7.639	−3.55	0.0006
Indicator for South Dakota	−70.827	9.011	−7.86	<0.0001
Indicator for Utah	64.321	8.721	7.38	<0.0001
Indicator for Virginia	−22.074	7.454	−2.96	0.0038
Indicator for Washington	32.040	12.270	2.61	0.0103
Indicator for Wisconsin	27.198	7.787	3.49	0.0007
Trend adjustor for Alabama	−3.333	1.769	−1.88	0.0622
Trend adjustor for California	3.555	1.765	2.01	0.0466
Trend adjustor for Illinois	2.645	1.201	2.2	0.0299
Trend adjustor for Maryland	4.453	1.144	3.89	0.0002
Trend adjustor for Nebraska	3.668	1.335	2.75	0.0071
Trend adjustor for North Dakota	3.960	1.754	2.26	0.0261
Trend adjustor for Oklahoma	4.724	1.758	2.69	0.0084
Trend adjustor for Texas	−3.853	1.313	−2.93	0.0041
Trend adjustor for Washington	−3.860	1.758	−2.2	0.0303

NOTES: $N = 144$; $R^2_{adj} = 0.93$; root MSE = 12.4 gpcd; mean APE = 6.3%.

$LG_{it} =$ indicator for state groundwater law system (equals 1 if prior appropriation, 0 otherwise)

$LS_{it} =$ indicator for state surface water law system (equals 1 if prior appropriation, 0 otherwise)

$a_i =$ intercept adjustor for individual states

$S_i =$ indicator for individual states (equals 1 if the state is included in the
model, 0 otherwise)

$b_i =$ trend coefficient describing changes in withdrawals in gpcd per year for
individual states

$Y_i =$ year since 1980 (equals 5 for 1985, 10 for 1990, and 15 for 1995)

$D_i =$ indicator for state-specific trend (equals 1 gpd if the state is included in
the model, 0 gpd otherwise)

This model in Table 6.2, which contains significant "intercept effects" for 23 individual states and trend effects for 9 states, explained 93 percent of variance in per capita withdrawals in the 1980–1990 data. The removal of one data year (1995) and the addition of binary variables had some effect on the estimated coefficients of the continuous variables when compared to those presented in Table 6.1. The coefficients of the price and precipitation variables have significantly less negative values when compared to the explanatory model in Table 6.1. The differences in the estimated coefficients indicate that the structural component of the model in Table 6.1 is not robust with respect to changes in the number of observations in the data and the inclusion of the binary variables to designate individual states. However, all six coefficients (including the binary water rights indicator variables) in Table 6.2 have the expected signs and remain statistically significant.

The model statistics shown below Table 6.2 indicate that the mean absolute percentage error (APE) for in-sample predictions is 6.3 percent as compared to 12.9 percent in the explanatory model (Table 6.1). The out-of-sample prediction errors for the 1995 data, which were not used to estimate the model, are shown for individual states in Table 6.3.

The comparison of the predicted and actual values in Table 6.3 indicates that the predictions for the 1995 data year were within ±10 percent for 24 states. In 17 states, the 1995 predictions were between ±10 percent and ±20 percent, and in 8 states, the absolute percentage error was greater than 20 percent. The largest error of 33.5 percent was obtained for California. The mean absolute percentage error for all 48 states in 1995 was 13.4 percent. The mean APE of 13.4 percent would also apply to the estimates of total public supply withdrawals for each of the lower 48 states (in million gallons per day), generated by multiplying the estimated per capita value by population served. If the model predictions for individual states were to be used to prepare an estimate of the total national public supply withdrawals for 1995, then due to the compensating positive and negative

TABLE 6.3 "Out-of-Sample" Predictions of Per Capita Public Supply Withdrawals for 1995

State	Withdrawals (mgd)			State	Withdrawals (mgd)		
	Actual	Predicted	% Diff.		Actual	Predicted	% Diff.
Alabama	237.1	171.4	−27.7	Nebraska	221.4	272.9	23.3
Arizona	206.1	231.5	12.3	Nevada	324.8	339.8	4.6
Arkansas	190.8	171.3	−10.2	New Hampshire	140.0	154.5	10.4
California	184.5	246.4	33.5	New Jersey	149.5	168.8	12.9
Colorado	207.7	238.5	14.8	New Mexico	225.4	239.7	6.3
Connecticut	155.2	164.0	5.7	New York	185.1	188.6	1.9
Delaware	158.6	169.9	7.1	North Carolina	162.1	170.1	4.9
Florida	169.1	172.0	1.7	North Dakota	148.9	176.9	18.7
Georgia	195.5	177.2	−9.3	Ohio	153.1	174.8	14.2
Idaho	242.9	256.7	5.7	Oklahoma	193.8	210.5	8.6
Illinois	175.3	217.7	24.2	Oregon	234.8	213.0	−9.3
Indiana	156.1	169.2	8.4	Pennsylvania	170.8	204.0	19.5
Iowa	173.2	171.1	−1.2	Rhode Island	130.2	147.1	13.0
Kansas	159.1	157.8	−0.8	South Carolina	199.6	158.8	−20.4
Kentucky	147.8	163.5	10.6	South Dakota	146.7	158.6	8.1
Louisiana	165.8	175.8	6.0	Tennessee	175.9	166.4	−5.4
Maine	141.7	160.5	13.3	Texas	187.7	169.0	−9.9
Maryland	200.0	244.7	22.3	Utah	268.9	304.2	13.1
Massachusetts	130.0	160.5	23.5	Vermont	148.3	164.2	10.7
Michigan	188.4	183.8	−2.4	Virginia	158.5	155.0	−2.2
Minnesota	145.2	178.0	22.6	Washington	266.3	216.0	−18.9
Mississippi	151.8	158.1	4.1	West Virginia	133.7	149.2	11.6
Missouri	161.5	167.9	4.0	Wisconsin	168.6	195.4	15.9
Montana	222.1	253.6	14.2	Wyoming	260.6	250.7	−3.8

prediction errors among individual states, the prediction error in the national total would be +2.2 percent.

STATE-LEVEL MODELS FOR THERMOELECTRIC WITHDRAWALS

State-level data for public water supply withdrawals are more accurate than data for thermoelectric cooling withdrawals. This is because public supply withdrawals are generally metered while withdrawals for thermoelectric cooling are more likely to be estimated based on pumping times and rated capacities of pumps.

The largest quantity of withdrawals from surface (and groundwater) sources is for thermoelectric power. The variables that can be examined as potential predictors of state-level thermoelectric withdrawals include the following:

- *Energy generation by fuel type:* total thermoelectric generation, percent coal generation, percent petroleum generation, percent natural gas generation, and percent nuclear generation;
- *Generation by method:* percent nuclear steam generation, percent conventional steam, and percent internal combustion;
- *Installed generation capacity:* total generation capacity (megawatt), percent conventional steam, percent nuclear steam, and percent internal combustion;
- *Availability of cooling towers:* total number of cooling towers, rated generation capacity with cooling towers (megawatt), number of cooling towers at coal steam plants, capacity (coal) with cooling towers (megawatt), number of cooling towers at petroleum/gas plants, capacity with cooling towers at petroleum/gas steam plants (megawatt);
- *Weather conditions:* cooling degree-days, heating degree-days, average annual air temperature;
- *State water law:* prior appropriation, riparian, riparian with permits; and
- *Number of generating units:* within coal, petroleum, gas, and nuclear categories.

Total withdrawals for thermoelectric power differ greatly among states, and the reported volumes are not well correlated with the total amount of thermoelectric generation in each state. However, when states with small generation and low water withdrawals (i.e., generally less than 1,000 MGD) are removed from the sample, a significant improvement in this relationship is achieved.

Table 6.4 presents a multivariate model of unit water withdrawals expressed as gallons per kilowatt hour for a group of states with large generation. The estimated regression coefficients indicate that the best explanatory variable for the quantity of withdrawals per kilowatt hour is percent generation capacity in plants that utilize "closed-loop" systems (i.e., cooling towers) relative to capacity

TABLE 6.4 Linear Model of Thermoelectric Withdrawals per Kilowatt-Hour

Variable	Estimated Withdrawal	*t* Ratio	Prob. >|*t*|
Intercept	49.376	15.53	<0.0001
Percent generation capacity with cooling towers	−0.362	−8.02	<0.0001
Percent utilization of existing capacity	−0.423	−4.99	<0.0001
Percent generation from coal	−0.096	−3.43	0.0009
Average size of generating units	0.174	6.34	<0.0001
Total heating degree-days	0.002	4.11	<0.0001
States w/ prior appropr. surface water law	3.962	−2.9	0.0047

NOTES: $N = 91$, $R2 = 0.80$; root MSE = 6.3 gal./kWh; mean APE = 17.6%.

of plants that depend on "once-through" cooling systems. Other predictors include percent utilization of existing capacity, percent thermoelectric generation from coal fuel, average size of generating units, and total heating degree-days. Additional explanation is provided by the "water law" variable, which indicates lower unit water withdrawals in states with prior appropriation surface water law (primarily western states). The model reveals the underlying structure of the thermoelectric demand despite the high level of data aggregation. All model coefficients have the expected signs and are statistically significant. They point to the importance of technological alternatives (i.e., once-through vs. evaporative cooling or combined-cycle generation) as determinants of water withdrawals.

Although the regression model in Table 6.4 explains 80 percent of the variance in per kilowatt-hour thermoelectric water withdrawals, the mean absolute percentage error for in-sample predictions remains relatively high at 17.6 percent. As in the public supply sector, improved predictions of the thermoelectric withdrawals model could be obtained by introducing binary state indicator variables.

Potential Model Improvements

The first step in improving the predictive properties of regression models of water use would be to enhance the quality of the data used in estimating the model parameters. Indeed, one of the advantages to regression approaches is that they may reveal cause-effect relationships that provide insight into data limitations. That is, because errors in the explanatory variables can be minimized, poor model predictions for individual states or years may suggest data errors in the USGS water use compilations. Thus, this approach may add value to both the assessment of water use and the quality control of the data. The effort expended to improve the data must, of course, be balanced with the effort expended to obtain reliable prediction variables.

Historical and current data on some of these explanatory variables exist, as they are routinely collected and archived by federal, state, and local governmental agencies. For example, the NWUIP currently collects data on population served and irrigated acreage. However, data on other variables, such as retail and wholesale water prices and thermoelectric generation capacity with cooling towers, are not routinely collected. If justified by their explanatory contribution in water use estimation models, such data collection and archiving could be added to NWUIP or state-level programs.

A second step would involve respecification of the predictive models. The relationships between the independent and dependent variables are likely to be different between the states of the humid East and the more arid West. The states, therefore, could be separated into groups based on geography and separate relationships estimated for groups of states, thus allowing the regression coefficients to vary among different regions of the country.

A third step would involve the introduction of additional variables in the multivariate regressions. Such variables, like marginal price of water or water conservation activity, are difficult to measure at the state level although they are known to have a significant influence on water use. For example, the results in Table 6.3 show a significant overprediction of per capita rates in California, a state with an aggressive water conservation programs. A variable that could capture the differences in water conservation efforts through time and among the different states could potentially improve these predictions.

Also, developing multiple regression models of withdrawals at the county level and obtaining the state totals by summing up the county-level estimates could also improve the state-level estimates of water withdrawals. However, the county-level data, which were developed by the NWUIP for 1985, 1990, and 1995, contain many apparent errors, and reliable models can be developed only after the accuracy of a number of data points can be verified.

Finally, given the potential for improvements in the data and models through the application of the "science of water use," the final statistical models for estimating water use may be of different form and structure than the examples developed here. However, the linear models used in this chapter to illustrate the approach do show the promise of the method.

CONCLUSIONS AND RECOMMENDATIONS

The examples presented in this chapter indicate that statistical models are a promising approach for estimating some categories of water withdrawals per unit (i.e., per capita or per kilowatt hour) within an acceptable estimation error. Based on the results presented in this chapter, the following conclusions can be drawn:

• A large number of potential explanatory variables for water use exist and can be used in constructing multiple regression models for the major categories of water withdrawals.

• Despite the state-level aggregation of the withdrawal data, these regression models reveal the underlying structure of water demand within several major sectors of use, and they reveal the key explanatory variables.

• The predictive properties of the models can be improved through appropriately specified models and through the inclusion of both the standard explanatory variables and the indicator variables for individual states or counties to capture their "unique" water use characteristics as well as state-specific trends in usage rates over time.

• The coefficients derived from regression models for adjustment of water use according to weather variations may be helpful in adjusting state-level water use estimates developed through statistical sampling or other means for departures from normal weather conditions in the year the estimates were made.

In summary, the data on water withdrawals and use that have accumulated under the NWUIP offer an excellent opportunity for advancing the "science of water use" and for understanding the structure and trends in national water use. The development of statistical models can be helpful in the quality assurance/ quality control process for future national compilations and for estimating water use in states or counties with inadequate data on withdrawals. Still, many challenges relating to data quality, inconsistent variable definitions, and statistical methodology need to be addressed, and they represent a fertile area for applied research as part of the NWUIP. *As part of its research on estimation methods, the USGS should undertake a systematic investigation of water use models as it has done for estimation of river loads, urban nonpoint pollution discharges, and other hydrologic quantities.*

7

A Vision for the NWUIP

The National Water-Use Information Program (NWUIP) currently focuses on the collection, estimation, and management of county-level water use data. The committee found significant opportunities to advance the techniques and technology used to estimate and manage national water use data. There is a compelling need for unbiased, science-based national water use information— information that will only become more vital for management and policy decisions in the future. Together, these findings led the committee to frame a broad vision for the NWUIP.

Long-term planning and management decisions need water use information that shows how water is an essential economic commodity and a vital natural resource. To use water use information in investigations of water resources or ecosystems requires placing water use prominently within the hydrologic cycle. The NWUIP should, therefore, be viewed as much more than a data collection and database management program focused on county-level categorical water use. Rather, it should become more integrated with the other efforts of the U.S. Geological Survey (USGS) to provide unbiased water resources information for assessing impacts and sustainability of current and future water use.

REQUIREMENTS FOR ANY NATIONAL
PROGRAM OF WATER USE ESTIMATION

First and foremost, the NWUIP needs to obtain estimates of water use that meet unambiguous and meaningful objectives. Setting clear objectives is essential to selecting the estimation methods at the proper resolution and scale. Although many possible objectives can be chosen, most of these should be at the national level.

The following are proposed as minimum requirements for any national program for accurate water use estimation:

• Water use estimates should be consistent across the country and should support meaningful comparisons of patterns and trends across geographic, climatic, and political boundaries.
• Estimates should be linked to the surface and groundwater resources affected by withdrawals. Linking water use to the resource emphasizes the importance of such water use categories as consumptive use, recharge, and return flows in maintaining water resources.
• Estimates should be consistent with the categories and spatial structure of previous national estimates, enabling the analysis of regional and national water use trends.
• Fundamental changes in the behavior and technologies that determine water use patterns should be detected and quantified. These changes (in, for example, water use efficiency, application rates, sources, consumptive use, etc.) may be caused by institutional changes (such as the legal *need* to exercise water rights), macroeconomic changes, or hydroclimatic changes. This suggests that before each national assessment is done, methods have to be developed to screen or "pre-sample" in order to identify major changes in water use and to refine sampling and estimation methods.
• Statistical sampling and inference should be done within the context of an unambiguous and statistically meaningful definition of the *populations* of water users to be estimated. Only by this kind of analysis can a hypothesis-based framework for estimation, error analysis, and the analytical determination of sample size requirements be meaningfully done.
• Estimation and data collection should incorporate error analysis designed to give relative standard errors (and thus confidence limits) on *all* quantities, including estimated changes from prior estimates. Error analysis should distinguish different types and sources of error (e.g., sample vs. nonsample error, nonresponse bias, etc.).
• Estimates should account for "intermittent" changes in water use (e.g., irrigation in the humid East only in dry years; the exercise of junior water rights in the West only in wet years).
• Estimates should account for interbasin transfers (e.g., Colorado River diversions to southern California) and for transfers between surface and groundwater sources (e.g., aquifer storage and recovery) over multiple scales.

GOALS OF THE NWUIP

By meeting these requirements, the NWUIP would address two complementary goals: (1) understanding *spatial* water use, and (2) understanding *hydrologic* water use. *Spatial* water use refers to estimates of the total use of surface and groundwater, by category, within any designated area. The area may be defined by political or hydrologic boundaries. Understanding spatial water use requires knowledge of human behavior and consumption patterns and decisions. It directly supports information generation associated with water use as a material flow in regional and national economic activities. *Hydrologic* water use refers to water use as the human component of the hydrologic cycle and emphasizes the impact of humans on the natural resource. Understanding hydrologic water use requires determining the magnitude, location, and timing of water withdrawals, and determining consumptive use and return flows affecting a designated resource. Although consumptive use and return flows are essential for understanding the water budget, these water uses are often calculated as the residual in a hydrologic budget and are not routinely obtained from site-specific water withdrawal records.

The NWUIP can and should characterize and quantify the importance of water as an essential commodity in the economy and the use of water as a critical hydrologic stress affecting the sustainability of the nation's water resources. This requires applied research and techniques development in both estimation and sampling techniques as well as science-based research on the determinants and impacts of water use behavior. Such a vision for the NWUIP leads to the USGS fully integrating water use with process-based science in its hydrologic and water resources investigations.

CONCEPTUAL FRAMEWORK FOR THE NWUIP

A conceptual framework for the NWUIP that addresses the goals identified above is illustrated in Figure 7.1. As suggested in the figure and in this report, a future NWUIP would be supported by a foundation of water use *data*, water use *estimation*, and water use *science*.

Water use data refers to measurements or estimates of the amount of water use at a site or for a region. These data would include direct measurement, as well as estimates obtained from surrogate measurements of activities involving water use. The term also refers to national geospatial databases and data inventories on water resources and water facilities, which support water use estimation.

Water use estimation refers to statistical sampling, inference, and estimation techniques for estimating spatial water use. An array of estimation techniques will be needed for the NWUIP, since water use estimation is inherently tied to the water use data available for individual states and the nation, and the goals of the NWUIP. With the growing water use data that are available, there are significant opportunities for advancing water use estimation.

Water use science refers to the hypothesis-driven investigation of the behavior and phenomena that determine spatial and temporal patterns of water use. This science will directly contribute to the development of techniques that support improved water use estimation. Water use science, discussed further in Chapter 8, also includes scientific assessment of the sustainability and impacts of

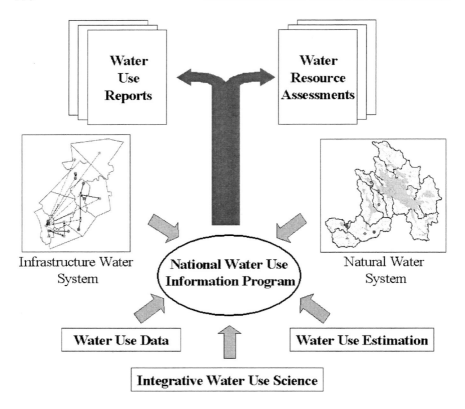

FIGURE 7.1 Conceptual framework for the National Water-Use Information Program.

water use on aquatic ecosystems, on the hydrologic cycle, and on the reliability and vulnerability of the nation's water resources.

The goals of understanding spatial water use and hydrologic water use require a view of water use from the perspective of both the *infrastructure water system* and the *natural water system*. The infrastructure water system is described by locations of water withdrawals and the movement of water through the landscape in constructed water systems to users and eventually to points of discharge. The natural water system is described by streams, rivers, lakes, aquifers, and watersheds, which coexist with the infrastructure water system. Exchanges of water occur between the two systems, primarily at the points of water withdrawal and discharge.

The conceptual framework envisages two major applications of information and data provided by the NWUIP:

• *Water use reports*, incorporating the results of spatial water use estimation, currently being done by the USGS in its five-year national summaries of water use, and also incorporating interpretive summaries of water use data prepared at the state and regional levels.

• *Water resource assessments*, involving hydrologic water use analysis, where the availability of water from surface and groundwater resources is balanced against how these waters are being used.

The sections that follow elucidate some of the major features on the conceptual framework for the NWUIP.

THE NATURAL AND INFRASTRUCTURE WATER SYSTEMS

Site-specific water use data (represented as points and/or polygons) can be viewed as data points in a spatial network, wherein each water use is associated with a node, and connections to other water use nodes are described by a set of directed links. Such a *link-node* representation gives a sense of direction to the flow of water use, just as the National Hydrography Dataset defines upstream-downstream relationships in the stream network. A link-node data structure can be used in a spatial database, defining all of the upstream uses that affect each water use node, as well as all of the downstream uses that may be affected by every water use.

A network data structure for water use provides additional information, linking the impacts of water use in both the natural and the built environments. Links to the natural water system directly connect water use to surface and groundwater sources by withdrawals, discharges, recharge, instream flow, and ecosystem impacts. Links to the human environment connect water use to the infrastructure water system of pipes, canals, treatment plants, and other man-made structures that extract, convey, distribute, and collect water. The natural water system, at least with respect to surface water, is *convergent;* natural stream and river systems form dendritic networks draining many sources to a few sinks. The infrastructure water system can be *divergent*; a single source such as a reservoir may supply many points of water use distributed over the landscape.

Conceptually, a link-node data structure for water use integrates spatially referenced water use in the natural water system and the infrastructure water system. A region in central New Hampshire near Lake Winnipesaukee is shown in Figure 7.2, for which the link-node water use data were obtained from the New England Water-Use Data System (NEWUDS) (Horne, 2001). Figure 7.2(a) shows the natural water system for this region, composed of Lake Winnipesaukee and its neighboring lakes, the streams and watersheds of this region, and groundwater aquifers (not shown). Superimposed on the natural water system are the points of withdrawal of water from groundwater and surface water, and a point of discharge of water back to the natural water system.

(a)

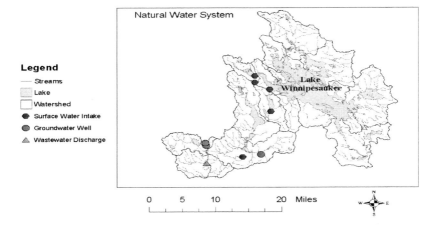

(b)

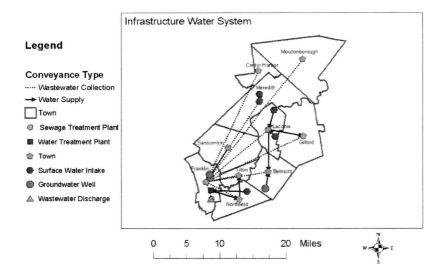

FIGURE 7.2 (a) Natural water system near Lake Winnipesaukee in central New Hampshire, (b) infrastructure water system for the towns in this region, and (c) combination of the natural and infrastructure water systems.

(c)

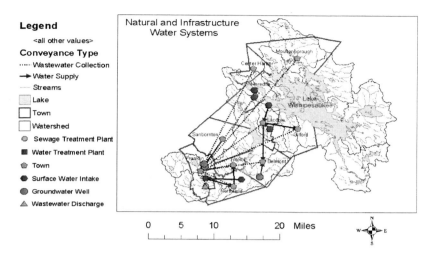

Legend

<all other values>

Conveyance Type

····· Wastewater Collection

➡ Water Supply

───── Streams

▢ Lake

▢ Town

▢ Watershed

◎ Sewage Treatment Plant

▧ Water Treatment Plant

♜ Town

● Surface Water Intake

◉ Groundwater Well

△ Wastewater Discharge

FIGURE 7.2 Continued

Figure 7.2(b) shows the infrastructure water system, composed of areas defining the legal boundaries of the towns, nodes at the centroids of these areas, and links symbolizing water movement within the infrastructure water system. The linkages are shown by solid lines for the water supply and distribution systems and by dashed lines for the wastewater collection and disposal systems. Although there are many points of water withdrawal, the local concern about wastewater discharge to lakes led to the construction of a regional wastewater treatment plant near the most downstream town, Franklin, from which the treated wastewater is discharged to the Merrimack River. Figure 7.2(c) shows the overlay of the natural and infrastructure water systems, from which the complexity of the interaction of these systems is readily apparent.

Link-node data structures retain all the attributes of a site-specific database, and they incorporate additional information showing how water use connects to the natural and infrastructure water systems. This information might include estimates of the leakage from the distribution system to the resource, infiltration from the resource into the wastewater collection system, consumptive use, and other water fluxes between the natural and infrastructure water systems. A network presentation of water use with natural hydrologic water movement enables mass balance calculations to be done for both systems. Indeed, an approximate mass balance inherently supports the hypothesis that the quality of data used to estimate water use (and other hydrologic fluxes) is high.

Linking each water use node to upstream influences and downstream effects provides a direct framework for examining policy and management questions, such as the impacts of water use on instream flow or the identification of source areas and vulnerable uses affected by a contaminant spill. Moreover, the network representation powerfully supports modeling and assessment of policy options for complex water-resource systems. Network flow models are robust tools used to plan and analyze extremely complex regional water-resource systems. For example, the U.S. Bureau of Reclamation (2000) used a network flow model to analyze the effects of diverting up to 1.4 million acre-feet from irrigation water use to instream flow needs in the Snake, Boise, and Payette River basins. Similarly, the CALVIN model (Howitt et al., 1999) is an extremely detailed network flow optimization model of the water resource system for the entire state of California. The model includes the statewide water system, surface and ground-water resources, storage and conveyance facilities, and agricultural, environmental, and urban water uses. It was developed to support operational planning and policy analysis of the economics and sustainability of the state's future water uses and water supplies (Newlin et al., 2001).

A link-node structure for water use data is a direct extension of a site-specific water use database. This conceptual framework for water use data, currently used in NEWUDS, can support the representation of water use in both the natural and infrastructure water systems. A link-node structure for water use data integrates the goals of spatial and hydrologic water use estimation conceived for the NWUIP, as suggested in Figure 7.1. The complementary characteristics of spatial and hydrologic water use are shown in Table 7.1.

TABLE 7.1 Characteristics of Spatial and Hydrologic Water Use

Spatial Water Use	Hydrologic Water Use
Estimates the quantity and type of water use in an arbitrarily defined area	Estimates spatial and temporal impacts of water use in the hydrologic cycle
Reflects water use as a material flow or as a commodity input to human activities	Reflects water use as a hydrologic flux resulting from human activities
Takes place principally within the infrastructure water system, but is connected to the hydrologic system through withdrawal, discharge, recharge, and return flows	Takes place principally within the natural water system, but is connected to the infrastructure water system through withdrawal, discharge, recharge, and return flows
NWUIP products describe quantity, trends, and the spatial and temporal distribution of water use, by water use category	NWUIP products describe impacts on vulnerability and sustainability of the resource, aquatic ecosystems, and riparian vegetation, by hydroclimatic category

A HIERARCHICAL WATER USE DATA STRUCTURE

The link-node data structure we suggest can also be visualized in the context of a data hierarchy depicted conceptually in Figure 7.3. At the lowest level, the focus is simply on water withdrawal site locations. Information from the federal databases described in Chapter 3 can be used to help develop an initial baseline water-withdrawal database for a state, if that information does not already exist. *When combined with appropriately selected estimation methods, these water withdrawal data can support spatial estimation of total water use*, such as that reported in the NWUIP's current five-year national summary reports.

At the second level in the hierarchy shown in Figure 7.3, water discharge points may be added to the water withdrawal points and linked by their location to the streams, rivers, lakes, and aquifers of the natural water system. If the amount of water withdrawn or discharged at each of these water use sites is known, *this information may be sufficient to support regional studies of water availability*.

At the third level in the hierarchy, information is added describing the water infrastructure: the jurisdictional boundaries of the institutions that manage water use, and the locations of their water and wastewater treatment plants and water distribution and collection systems. The quantity and quality of water flowing

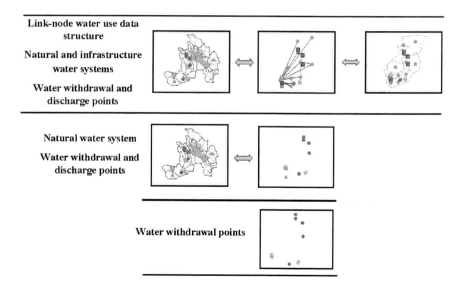

FIGURE 7.3 A hierarchy of water use representations. The symbols used are the same as those used in Figure 7.2.

through these facilities may also be defined. *A water use representation at this level can be used to create a comprehensive understanding of the movement of water through the landscape in both natural and infrastructure water systems, and to show how human use of water impacts the quantity, quality, and sustainability of water resource systems.* This level of complexity may be especially appropriate for jurisdictions with complex water management and accounting concerns, such as aquifer storage and recovery, artificial recharge, water reuse, desalination, total maximum daily load (TMDL) issues, and/or interbasin transfers.

Water use data, water use estimation, and water use science are the foundation for achieving the goals of understanding spatial and hydrologic water use within the natural and infrastructure water systems. The following sections outline these three components of a future NWUIP.

Water Use Data

The Arkansas water use program has an excellent database of permitted water withdrawal points maintained to support the state's regulatory responsibilities. To varying degrees, similar management and regulatory needs have resulted in comparable programs in many of the states. Concurrently, the maturation of geographic information system (GIS) technology as a standard data-management tool has made it common for states to organize and manage their water use data as GIS databases. The status and data availability of state water use programs have been summarized by the USGS district water use specialists (see Chapter 2).

The convergence of states' regulatory needs and the use of GIS technology has resulted in the emergence of *site-specific water use databases.* Many of these databases, including the Arkansas database with nearly 45,000 permitted water withdrawal points used throughout this report, are easily accessed and manipulated with over-the-counter GIS programs. Thus, it is now practical and efficient for the USGS to manage site-specific water use data (which would consist of literally millions of water withdrawal points and their associated attributes) for every state. The increasingly common availability of site-specific water use databases and GIS technology among the states is a key opportunity for the NWUIP.

The last few decades have also seen the emergence of major national databases on water-using facilities and activities, including many with site-specific data. Examples include the Safe Drinking Water Information System (SDWIS) database on public water supplies from the U.S. Environmental Protection Agency, databases on energy generation facilities and electricity production time series from the Energy Information Administration, and the Census of Agriculture information on irrigation water use from the U.S. Department of Agriculture (see Chapter 3). These databases are a rich source of basic information for new water use estimation techniques.

Although site-specific water-use databases are very useful, they have limitations. Total water use by category is easily calculated by summing water use for all withdrawal points within any geographically defined region (e.g., Arkansas). If the only goal for the NWUIP were to accurately estimate total withdrawals by county and state, a site-specific database would suffice. However, even detailed site-specific databases of permitted withdrawals alone cannot fully satisfy the requirements for national water use estimation proposed in this chapter. For example, site-specific data developed to support regulatory activities may omit water use that falls outside of the state's regulatory authority. This error may be negligible if the data are used to estimate total water use in a state, but may be unacceptably large for assessing drought vulnerability, sourcewater protection, or ecosystem impacts (see Box 7.1).

BOX 7.1
Limitations of Site-Specific Databases—Arkansas Example

Arkansas's site-specific water use database (see Chapter 3) is distinguished by its high-quality and complete coverage of permitted withdrawals. Global Positioning System (GPS) verification and georeferencing of water withdrawal points, and quality assurance/quality control activities that discovered and corrected errors in about 80% of the records, make this database a model for state regulatory programs.

However, alone, this comprehensive database is insufficient to satisfy the requirements for the NWUIP envisioned by the committee. Consider the site-specific data for domestic water use (Table 5.1). Only four of the nearly 45,000 water withdrawal points represent domestic water use. This shows not the absence of domestic water use in Arkansas, but rather the sparse number of domestic users large enough to be covered by the state's permit program. The USGS estimates domestic water use in Arkansas based on the difference between the state population and the population served by public supplies. Annual domestic water use not included in the state permit system can be estimated as:

450,000 persons × 89 gal./person/day × 365 days/year = 14.6 MG/year.

This represents only 0.1 percent of the total annual use reported in the state database. The error introduced by neglecting these domestic users is therefore negligible for estimating total water use for the state. However, this represents the primary water supply for about 20 percent of the population of Arkansas. Neglecting these domestic users would be unacceptable for assessment of drought vulnerability and statewide drought planning. Even the most accurate and complete site-specific water use database must be supplemented with indirect estimates for uses that are not captured by a regulatory permit program or, like instream flow, are not naturally associated with a single point of use.

Also, at a *regional* scale, spatial water use can be estimated without *any* site-specific data, and these estimates may be sufficient to meet many of the data-assessment requirements of the NWUIP. For example, Vörösmarty et al. (2000) estimated the global-scale vulnerability of the world's freshwater resources to climate change and increases in water use. They used publicly available spatial data to estimate global water use for 0.5-degree grid cells, and linked these estimates to the natural hydrologic drainage network extracted from digital elevation model data. To estimate urban water use without site-specific data, urban population was distributed between 1-km-resolution city polygons and 1-km grid cells classified as "city lights" from nighttime remote sensing images. The Vörösmarty et al. (2000) comparison of spatial water use estimates and regional water balances foreshadows the committee's vision of NWUIP integrating water use information in water resource assessments.

Site-specific permit data are typically associated with water withdrawal points, whereas water use is often more appropriately associated with an area (e.g., an irrigated field, a polygon describing the limits of a municipal water distribution system). Spatially distributed water use can be represented as a single point, such as the centroid of the population served or of a service area, the location of a treatment plant, or even the withdrawal location. However, this introduces estimation error whenever the arbitrary polygons used in estimation (e.g., census tracts, cities, counties, hydrologic units) do not correspond to the true spatial limits of each water use (see Box 7.2).

One of the strengths of GIS technology is that GIS data structures can combine both point and spatial data. Even so, no single spatial structure (e.g., polygons representing census tracts, county boundaries, or irrigated acreage) can fully support the goal of water use estimation for any arbitrarily defined geographic area. This goal also requires techniques to interpolate and apportion both water use data and supporting data such as population, employment, and irrigated acreage that are uniformly available from consistent national databases. These data are typically compiled by state, county, or census tract boundaries and therefore must be extrapolated over different spatial domains—such as a watershed. The extrapolation error from, for example, translating county-level data to a smaller hydrologic unit is inescapably part of national assessment programs such as the NWUIP. Identifying the best methods to quantify and minimize these approximation errors is a fruitful and widely applicable research area for the NWUIP.

Water Use Estimation

The nature and availability of data resources for water use estimation impose many constraints. For instance, the quantity and quality of water use data vary across the country, as each state tailors water use data collection to meet individual regulatory or resource assessment needs. As the discussion in subsequent sections will show, even in states with extensive site-specific water use and

---

BOX 7.2
Limitations of Site-Specific Databases—Maryland Example

To manage water use in a site-specific database, spatially distributed water uses can be represented by a single point such as the point of withdrawal or the location of a treatment plant. Consider public supply water use for the western Maryland city of Cumberland, in Allegany County. The city draws its raw water supply from surface water reservoirs in Bedford County, Pennsylvania. Treated disinfected potable water is distributed to the city's retail customers as well as to wholesale customers that operate their own distribution systems—including Allegany County. One of the city's largest single customers (accounting for about 25 percent of total demand) is a co-generation plant that uses potable "public supply" water for evaporative cooling in thermoelectric power generation. The city also sells treated disinfected drinking water to wholesale customers across the Potomac River in West Virginia. Depending on the convention adopted, representing this spatially distributed public supply system by a single point could variously shift the associated water use between three different counties in three different states and at least two separate categories of use.

Despite this complexity, the city of Cumberland's water use is less than 1 percent of the total public supply for the state of Maryland, which is dominated by public systems serving the Baltimore-Washington corridor. Errors introduced by treating Cumberland as a single water use point would have a small impact on the accuracy of estimated public supply water use for the state of Maryland. However, as with the Arkansas example in Box 7.1, these errors would have a significant impact in assessments of drought vulnerability, stresses on aquatic habitats, or the sustainability of water use in the affected watersheds.

---

spatial databases, there are limitations on the use of these databases in water use estimation. As a result, a "one size fits all" approach to water use estimation is impractical and undesirable for the NWUIP. Instead, a variety of estimation techniques, tailored to the data resources of individual states, are needed.

As the survey of estimation techniques in Chapter 4 showed, there are many promising approaches for water use estimation. Chapter 5 illustrated the advantages of statistical sampling of site-specific water use for estimating spatial water use and its uncertainty, while Chapter 6 demonstrated how statistical inference based on statewide categorical water use data can help reveal the structure of water use on a national scale. As part of the NWUIP, there is a strong need for the USGS to systematically compare these and other water use estimation methods in order to identify the techniques (or combination of techniques) best suited for the individual states or regions.

Although data resources supporting water use estimation are not consistent across the country, consistency is still important in the *products* from the NWUIP. For all states, water use estimates are needed for consistently defined sets of water use categories and spatial regions (e.g., counties). This situation makes the reporting of uncertainty in water use estimates critically important. Users of water use estimates need to be aware that uncertainties of estimates within these similar products are quite variable from one state to the next.

Water Use Science

Despite the importance of the conceptual framework outlined in this chapter, water use is more than a matter of databases, statistics, site locations, and estimates. One of the most fascinating aspects of water use is that human activity alters the quantity and quality of water and the patterns of water flow through the landscape. Human impacts on water systems have ecological consequences. Chapter 8 of the report looks at water use in this larger context of integrative water use science.

Still, with the growing availability of site-specific water use data and ancillary datasets, water use science also has an important role in improving water use estimation. There are many opportunities for research on technique development. These include the testing and evaluation of statistical sampling strategies, the transfer of information from data-rich areas for estimation in data-sparse regions, and the exploration of indirect techniques for water use estimation using ancillary datasets.

CONCLUSIONS AND RECOMMENDATIONS

There is a compelling national need for unbiased, science-based water use information for current and future management and policy decisions. To meet this need, the NWUIP must obtain estimates of water use that meet unambiguous and meaningful objectives. Specifically, the national program should at a minimum (1) produce consistent estimates across the country, (2) be linked to the surface and groundwater resources, (3) detect fundamental changes in the behavior and technologies that determine water use patterns, (4) utilize statistical sampling and inference within the context of an unambiguous and statistically meaningful definition of the *populations* of water users to be estimated, (5) incorporate error analysis designed to give relative standard errors, (6) account for "intermittent" changes in water use, and (7) account for interbasin transfers and transfers between surface and groundwater sources over multiple scales.

The committee sees the NWUIP addressing two complementary goals, namely, the understanding of (1) *spatial* water use, i.e., the use of surface and groundwater, by category, within any designated area, and (2) *hydrologic* water use, i.e., water use as the human component of the hydrologic cycle. Spatial water use is linked to human behavior and consumption patterns and decisions. Hydrologic water use emphasizes the impact of humans on the natural resource. *The NWUIP can and should characterize and quantify the importance of water use as both an essential commodity in the economy and water use as a critical hydrologic stress affecting the sustainability of the nation's water resources.* Achieving these goals requires applied research and techniques development in both estimation and sampling techniques as well as science-based research on the determinants and impacts of water use behavior.

These considerations suggest a conceptual framework for the NWUIP that links the *infrastructure water system*, described by locations of water withdrawals, water discharges, and the principal water facilities, with the *natural water system* of streams, rivers, lakes, aquifers, and watersheds. The increasingly common availability of site-specific water use databases and GIS technology among the states is a key enabling factor for this approach. Likewise, "link-node data structures" associate each water use with a node and connect the node to other water use nodes by a set of directed links. Such a link-node representation gives a sense of direction to the flow of water use.

Such a framework may not be appropriate at all scales (e.g., regional) and adequate for all purposes (e.g., drought management) and will need to be supplemented with other methods and frameworks. Regardless, any framework chosen should be supported by water use data, water use estimation, and water use science.

8

Integrative Water Use Science

Among the components of the hydrologic cycle, water use has unique qualities. Foremost among these is that water use has an important social science component. Water use science, therefore, must encompass and integrate both the social and natural sciences. We use the term "integrative water use science" to describe this multidisciplinary, hypothesis-driven investigation of the behaviors and phenomena that determine spatial and temporal patterns of water use, and of the impacts of water use on aquatic ecosystems, the hydrologic cycle, and the sustainability and vulnerability of the nation's water resources.

Water use is also an element of the geospatial data layer. In the past, the nation has not given water use data the importance it deserves; now, a reliable water use dataset can be developed with current technology. This was clearly illustrated in Chapters 3 and 7, which describe the emergence of site-specific water use databases and give many examples of how such a database, linked to a geographic information system (GIS), could illuminate the relations of water use to land use and hydrologic characteristics.

In this chapter, we expand upon the concept of integrative water use science, initially focusing on two very specific aspects of water use science that have not previously been addressed in the NWUIP: (1) the interrelationships among water use, water flow, land use, and water quality and (2) the estimation of instream flow for ecological needs. We argue that the NWUIP should consider expanding into these very important areas in collaboration with other U.S. Geological Survey (USGS) programs (e.g., National Water-Quality Assessment Program [NAWQA], National Streamflow Information Program [NSIP], and the Biological Resources Division), where the USGS can bring its many talents to bear upon a revitalized approach to water use science.

We conclude by presenting a holistic methodology for the NWUIP, primarily developed by Dr. Gregory E. Schwarz of the USGS, Branch of Systems Analysis. This approach would utilize the surface water quality model SPARROW (SPAtially Referenced Regressions On Watershed attributes) (Smith et al., 1997) to provide a framework for organizing and analyzing source data, a means for assessing data accuracy, and information on areas where data needs are most pressing.

SYNTHESIS OF WATER FLOW, WATER USE, LAND USE, AND WATER QUALITY

All uses of water affect its quality (i.e., its physical, chemical, and biological characteristics) (Getches et al., 1991). The quality of water determines what it can be used for; likewise, how water is used will change certain aspects of its quality. An example of the latter includes the increase in total dissolved solids (TDS) of irrigation water as it is diverted from a stream or reservoir or pumped from an aquifer, is applied to the land surface, undergoes evapotranspiration, leaches salt from the soil, and emerges as irrigation return flow. Water diverted for municipal consumption also generally suffers reduction in quality, as the water discharged from a wastewater treatment plant is generally "degraded" relative to the original diversion. Aquifer storage and recovery, an increasingly common water management practice, may lead to increased concentrations of arsenic, methyl mercury, and other metals (NRC, 2001). Indeed, Getches et al. (1991) stated that control of water use is one approach to protecting water quality and that most uncontrolled water-quality degradation relates to water uses authorized by state water allocation systems. Along these lines, it is interesting to note that return flow from irrigated agriculture, a major user of water, especially in the western United States, is specifically exempted as a nonpoint source of pollution under the Clean Water Act.

Water-quality degradation from use falls into four categories (Getches et al., 1991):

1. *Depletion degradation.* The consumption of water results in a higher concentration of pollutants because the remaining water is less able to dilute them (e.g., consumptive use of a stream diversion reduces the stream's ability to dilute pollution).

2. *Physical alteration.* Some uses of water directly alter the physical characteristics of the water (e.g., storage in a reservoir changes temperature, dissolved oxygen content, etc.).

3. *Pollution migration.* Water use causes preexisting pollution to contaminate other waters (e.g., pumping from an aquifer may induce the migration of

contaminated water into the aquifer; diversions from a stream may cause salt-water encroachment).

4. *Incidental pollution.* Water use causes pollutants to enter waterways other than from discrete point sources (e.g., irrigation water leaches salts from soils, causing the salts to enter surface waters and groundwaters as part of irrigation return flows).

Water use is also linked to land use. Different land uses will dictate different degrees of water use. Even within a given land use—agriculture, for example—water use can vary dramatically (see Table 3.1). Site-specific water use data-bases (see Chapter 7), such as the Arkansas database, will enable the Naitonal Water-Use Information Program (NWUIP) to assess the interrelationship between land use and water use.

The above discussion suggests that water use, land use, water flow, and water quality, although often considered separate from one another, are inextrica-bly linked. They should not be treated independently of the other, although this is often the case insofar as water allocations are concerned. Box 8.1 provides an illustrative example from one of the nation's most heavily allocated basins—the Colorado River basin of the western United States.

The current NWUIP deals with use and flows only. The usefulness of the NWUIP would be increased were it to consider water quality. The USGS could lead the way in the integration of land use, water use, water flow, and water quality data by expanding the current NWUIP. But given the budget constraints of the present, how can water quality be integrated into the NWUIP short of calling for an expensive program to monitor our nation's waters? Fortunately, there are a number of current programs with which the NWUIP might collabo-rate. A few of the more visible programs are described in the next section.

SOURCES OF WATER-QUALITY DATA FOR AN EXPANDED NWUIP

Water-quality data may come from many sources, both within and outside of the USGS. These sources include the USGS National Water Quality Assessment (NAWQA) and National Stream Quality Accounting Network (NASQAN) Pro-grams, the U.S. Environmental Protection Agency (EPA) the U.S. Bureau of Reclamation (USBR) and the U.S. Army Corps of Engineers (ACE).

The National Water-Quality Assessment Program

The NAWQA Program (http://water.usgs.gov/nawqa; see also Gilliom et al., 1995; Hirsch et al., 1988; Leahy et al., 1990) is an ambitious program of the USGS designed to describe the status and trends in the quality of the nation's groundwaters and surface waters and to understand the natural and anthropogenic

---

BOX 8.1
The Inseparability of Water Use, Land Use, and Water Quality:
The Colorado River Basin

The Colorado River of the western United States and Mexico is subject to salinity problems from both natural and anthropogenic sources. Salts derived from the Mancos Shale in the upper basin have created problems for irrigated agriculture in the Grand Valley of western Colorado for over 100 years (MacDonnell, 1999). Construction of substantial drainage systems around the beginning of the twentieth century did much to mitigate the local salinity problem by essentially transferring the salts downstream to the lower basin (MacDonnell, 1999). About 50 percent of the salts in the Colorado River at Hoover Dam (just outside Las Vegas) are derived from natural sources, whereas 37 percent are derived from irrigation (MacDonnell, 1999). Thus, salinity, induced in part by land use in the upper basin, has made the river's water less usable for municipal, domestic, and some industrial uses and has also limited the types and the yields of crops that can be grown in the lower basin (MacDonnell, 1999).

Salinity problems in the lower Colorado River have also limited Mexico's ability to use its allotment of Colorado River water and have strained relations between the two countries. In 1944, the United States and Mexico agreed that Mexico would receive an allotment of 1.5 million acre-feet of Colorado River water each year and up to 1.7 million acre-feet in water-surplus years (Garcia-Acevedo, 2001). It is important to note that the treaty made no mention of the *quality* of this water, an omission that would return to haunt Mexico about 20 years later. Mexico would use its allotment to further growth and irrigated agriculture in the Mexicali Valley. However, in 1961, the Wellton-Mohawk Irrigation District of Arizona began disposing of its agricultural wastewater in the Colorado River, thereby increasing the salinity of the river to over 2,000 parts per million (ppm) of total dissolved solids (TDS) at the Mexico border. The deleterious impact of this salinity increase on agriculture in the Mexicali Valley was swift and severe (Kliot et al., 1997), as agricultural lands were ruined (Garcia-Acevedo, 2001). Various agreements and projects since then (e.g., "Minute 242," signed in 1973) have ameliorated the salinity problem somewhat (it is now about 1,000 ppm), although the agricultural economy of the Mexicali Valley has never fully recovered from the salinity damage.

Mexico had indeed been receiving its allotment—the *quantity* was there, but the *quality,* lowered by the nature of the *land use* in the upper and lower basins, impaired the water for its intended *use,* irrigated agriculture. Although the water Mexico obtained from the Colorado River could not be used for irrigated agriculture, perhaps it could have been used for other purposes had the land in the Mexicali Valley been used for another purpose.

---

factors affecting their status. Water-quality investigations are being conducted in 59 "study units" (SUs), which cover about half the land area of the United States and about 60 percent to 70 percent of the nation's water use and population served by public water supplies.

　　　The NAWQA Program began as a pilot program of 7 study units in 1988. In 1991, the number of SUs was increased to 20, and the program reached its full complement of 60 SUs in 1997 (subsequently reduced to 59 by combining two adjacent SUs in New England). The SUs represent portions of major river basins and aquifer systems. Budget limitations will necessitate a cut in the number of SUs to 42 (this is the estimate as of August 2001) as the NAWQA Program enters its Cycle II (second 10 years) phase in 2001. Even with the planned reductions, the NAWQA Program will still cover about 40 percent of the nation's land area, comprising about 60 percent of its drinking water use. The NAWQA Program is far more than just a data collection/monitoring program, as it seeks to establish cause-and-effect relationships and quantify the effects of land use on water quality. It also has a "national synthesis" component; a report on nutrients and pesticides (USGS, 1999) has already been produced; the next reports will be on VOCs (volatile organic compounds) and trace elements. Because the NAWQA Program does not cover the entire nation, any of its water-quality information that would be used in an expanded NWUIP would have to be augmented by data from other programs.

　　　Although much of the existing water use data are organized by political unit, it may be feasible in some cases for the NAWQA study units themselves to serve as modeling units to develop the science behind water use estimates. Statistical estimation models for water use could be developed and tested. Sampling protocols could be examined where land use information, water withdrawal, and water discharge points are usually known. The NAWQA study units could effectively become water use investigation laboratories for developing the science behind the water use estimates.

　　　Although NAWQA activities alone could not provide the data required for a redesigned NWUIP, NAWQA's overall themes of water-quality status, trends, and understanding fit well with the NWUIP's objective of providing information on status of and trends of the nation's water use. The NAWQA Program was originally conceived to determine if the nation's water quality was improving and whether all the money that had been spent cleaning up the nation's waters had had any effect. It would be well worthwhile to link the NAWQA and NWUIP programs, as they both deal with important aspects of water—its use (and flows) and quality. An expanded NWUIP could provide a snapshot of the nation's water use and quality every five years, augmenting the numerous NAWQA reports and complementing NAWQA by assessing the relationships between land use and water use.

The National Stream Quality Accounting Network Program

　　　The USGS NASQAN Program has been in effect since 1973 (http://water.usgs.gov/nasqan/), collecting water-quality information at over 500 locations. Such a suite of data would fit nicely with water use data. However, since

1995, the program has been cut drastically and now monitors only 39 sites in four major river basins: the Columbia, Mississippi-Missouri, Colorado, and Rio Grande. Barring a restoration of this program, the current NASQAN Program could only provide a small set of streamflow quality measurements.

Environmental Protection Agency

The EPA is another organization concerned with water quality and has programs that could provide water-quality data to an expanded NWUIP. A number of EPA-sponsored programs collect water-quality data; only two will be mentioned here. The Environmental Monitoring and Assessment Program (EMAP) is monitoring and assessing water quality as part of its mission (http://www.epa.gov/emap/). The National Water Quality Inventory, submitted to each Congress under Section 305(b) of the Clean Water Act, contains a wealth of information on surface waters and groundwaters (U.S. EPA, 2000). The inventory is based upon data collected by the states and Indian tribes; these Section 305(b) quality data could be obtained by NWUIP and integrated into its reports. Because NWUIP already relies on state cooperation and data for its work, its use of additional state data would be appropriate.

U.S. Bureau of Reclamation and U.S. Army Corps of Engineers

Both the USBR and the ACE provide water for drinking, recreation, irrigation, and aquatic life, primarily through reservoirs they construct and operate. These agencies perform water-quality monitoring for a variety of uses: aquatic life, fish consumption, primary contact, secondary contact, drinking water supply, and agriculture.

The USBR is currently concerned with the quality of irrigation water supplied by its projects through its National Irrigation Water Quality Program (http://www.usbr.gov/niwqp/). Because irrigation return flow often empties into reservoirs—e.g., Kesterson Reservoir in the San Joaquin Valley, whose wildlife has suffered from selenium poisoning (NRC, 1989)—the USBR is monitoring quality in a number of its reservoirs. These USBR-collected data could be integrated into NWUIP's work.

Some of the ACE districts/regions have very active reservoir water-quality monitoring programs. The Northwestern Division of the Corps' Missouri River Region (Omaha and Kansas City Districts) has its Water Quality Management Program–Missouri River Region Lake Projects. The Omaha District alone is conducting water-quality studies in over 30 lakes and reservoirs. The NWUIP could use these data in its effort to characterize the nation's water quality.

Other Agencies

Numerous other agencies have water-quality data that may be made available for the NWUIP. Prominent among these are the U.S. Department of Agriculture (USDA), U.S. Forest Service, the National Park Service (NPS), the Bureau of Land Management (BLM), and the National Oceanic and Atmospheric Administration (NOAA).

Coordinating NWUIP and the NAWQA Program

An expanded (i.e., water use *and* water quality) NWUIP and the NAWQA Program would have much common ground for collaboration. The NAWQA Program realizes the interconnection between use (land and water) and quality in its effort to understand cause-and-effect vis-à-vis water quality. Further, it requires water use data: the NAWQA Program needs specific public supply and domestic water use information and will be affected by the elimination of some categories in the 2000 NWUIP report (M. Maupin, USGS, personal communication, 2000).

An excellent example of a NAWQA project where water use and water quality are interconnected and integrated is the Santa Ana Basin SU in Southern California (http://water.wr.usgs.gov/sana_nawqa/). Although this study unit is rather small (2,700 square miles) with a highly "engineered" hydrologic cycle and does have a number of agencies collecting both use and quality data, the interdependence of the two types of data is clearly indicated and appreciated. Perhaps this study unit could be used as a model for similar water use–water quality studies.

In addition, the SPARROW model (Smith et al., 1997), which has been used extensively in the NAWQA Program as a regional water-quality assessment tool, may also be useful in an expanded NWUIP. Although SPARROW was not developed under the aegis of the NAWQA Program, it has been implemented in a number of NAWQA studies. The relationship between SPARROW and NWUIP will be further explored at the end of this chapter.

Coordinating the two programs, or at least elements of them, makes sense. Both are concerned with status and trends. The NAWQA Program already has a national team in place, and it does have a demonstrated need for water use data. It has a national synthesis component and seeks to link land use and water quality. Its study units do not cover the entire country, but many USGS district offices have NAWQA teams.

ESTIMATING INSTREAM FLOW FOR ECOLOGICAL NEEDS

Estimation of the instream flows needed for environmental (or ecological) uses is becoming increasingly important as greater emphasis is placed on the importance of keeping water in streams for ecosystems (Gillilan and Brown, 1997). Development has put heavy demands upon streams, especially in the arid West, where already-scarce surface water is subject to many demands. Each additional diversion leaves less water in stream channels to satisfy a variety of ecological uses: fish and other aquatic organisms, wildlife, and riparian vegetation.

Changing Needs

Knowledge of how much water is used for instream ecological use will be critical, as there is growing awareness that the benefits of instream flow extend well beyond the immediate benefits to anglers and recreationists; this awareness is a powerful force changing traditional water management institutions (Thompson, 1999, p. 272). Leaving water in streams to protect aquatic and riparian ecosystems provides value to people who rarely, if ever, visit a stream, as well as to society as a whole (Gillilan and Brown, 1997). Environmental and recreational interests are increasingly pursuing an agenda that includes watershed restoration and protection of instream flows to restore and sustain a stream's historic ecological and hydrogeologic functions (Tarlock, 1999). A growing body of evidence indicates that the maintenance or restoration of whole riverine ecosystems requires a range of flow conditions, because different species have different flow optima and may be dependent on natural disturbances such as low and high flows (Poff et al., 1997; Sparks, 1992). Estimation of the range and timing of flows needed for instream ecological use is a major scientific challenge, but one that has been given increased significance by the Endangered Species Act, which in some cases could require mimicking of the natural flow regime to protect endangered organisms (e.g., Muth et al., 2000).

According to Richter et al. (1997), the growing need to predict the biological impacts (or recovery) associated with water management activities, and to set water management targets that maintain riverine biota and socially valuable goods and services associated with riverine ecosystems, has spawned a new approach to modeling instream flow needs. These newer models have as their primary aim the design of environmentally acceptable flow regimes (i.e., pattern of flow variation) to guide river management (Richter et al., 1997). Unfortunately, recent advances in understanding the relationship between hydrologic variability and ecosystem integrity have had minimal influence on the setting of instream flow requirements or on river ecosystem management (Richter et al., 1997).

Despite the interest in and importance of instream ecological flows, the most recent NWUIP report does not document instream uses except for hydroelectric power generation, which is relatively easy to quantify (Solley et al., 1998); the upcoming 2000 report will also follow this approach. Quantitative estimates for most instream uses, not just ecological ones, are difficult to compile on a national scale (Solley et al., 1998). However, because instream uses compete with offstream uses and affect water quality and water quantity, effective water resources planning and management dictate that such uses be individually quantified. California, for example, is documenting and calculating various instream uses, including certain kinds of environmental uses (Department of Water Resources, 1998).

Because the NWUIP relies upon state data for its national compilation, it is apparent that the program might have to calculate or estimate ecological instream uses for many states. With the relatively low level of funding currently provided to the NWUIP, this is not a realistic expectation. However, given the increasing importance of instream flows, the NWUIP should start exploring ways to estimate instream flows for ecological use for each state. It would be entirely appropriate for the NWUIP to develop techniques for estimating instream flow uses on a large-scale basis, perhaps by using resources within the USGS, such as the Midcontinent Ecological Science Center (MESC; http://www.mesc.usgs.gov).

Methods

One of the difficulties in estimating instream flows for ecological use is that different species have different instream flow needs. Riparian ecosystems require instream flows different in timing and magnitude from aquatic organisms, which often differ among themselves (e.g., fish vs. macroinvertebrates). As an example, throughout the West, inundation of alluvial floodplains is required to produce appropriate conditions for seed germination and seedling establishment of cottonwood trees, a major component of riparian forests. If floodplain inundation does not occur or if floodwaters recede too rapidly from the floodplain, cottonwood gallery forests decline (Auble et al., 1994; Rood and Mahoney, 1990).

Most methods to estimate environmental flows focus primarily on one or a few species that live in the wetted river channel. Thus, they cannot estimate instream flows for multipurpose ecological uses or for an entire ecosystem (NRC, 1992; Poff et al., 1997). Most of these methods are narrowly intended to establish minimum allowable flows. The simplest make use of easily analyzed flow data, of assumptions about the regional similarity of rivers, and of professional opinions of the minimal flow needs for selected fish species. Some of the more popular methods—e.g., Tennant; Modified Tennant; Habitat Quality Index (HQI); Wetted Perimeter; Aquatic Base Flow (ABF)—are briefly described and referenced in Lamb and Doerksen (1990) and Stalnaker et al. (1995).

A more sophisticated assessment of how changes in river flow affect aquatic habitat is contained in the Instream Flow Incremental Methodology (IFIM; Bovee and Milhous, 1978). The IFIM couples two models, one that describes the physical habitat preferences of fishes (occasionally macroinvertebrates) in terms of channel width, depth, velocity, and substrate, and a hydraulic model that estimates how available habitat space for fish life stages varies with discharge. The IFIM has been widely used as an organizational framework for formulating and evaluating alternative water management options related to production of one or a few fish species (Stalnaker et al., 1995). The IFIM has also been widely used for instream flow modeling (Lamb and Doerksen, 1990); however, in recent years it has been increasingly criticized for its lack of biological realism (Castleberry et al., 1996) and the accuracy of its physical simulations (Williams, 1996). The IFIM is labor- and data-intensive and requires field measurements and hydraulic modeling, and it is too costly and time-consuming for the NWUIP. Stalnaker et al. (1995) estimated that 80 percent of IFIM studies for a single stream reach might take 12 months and cost $45,000. It may be possible to extend IFIM techniques to the ecosystems planning and management realm, although it is too early to tell (Stalnaker et al., 1995).

Newer modeling techniques are "holistic" in that they incorporate the premise that environmental conditions that sustain the ecosystem will sustain the constituent individual species (Meyer et al., 1999). The details of species' responses to shifting conditions cannot be accurately modeled; however, the environmental regime can be. Thus, the guiding framework for these newer approaches is to describe the natural environmental regime usually in terms of the natural flow regime (e.g., Poff et al., 1997; Richter et al., 1996), although other environmental drivers such as temperature and sediment flux are also used. In this view, the integrity of riverine ecosystems varies in response to the deviation of the prevailing flow regime from the pre-impaired state. The natural regime is characterized in terms of the magnitude, frequency, duration, seasonal timing, and rate of change of flows—factors that are known to have demonstrable effects on aquatic habitat and ecological processes.

The NPS has developed an instream flow needs quantification known as "departure analysis" (Gillilan and Brown, 1997). This approach initially assumes that 100 percent of a stream's flow is required to maintain the natural riverine ecosystem. The method then incrementally subtracts—through modeling—small amounts of water until the reduced streamflow produces a quantifiable or observable impact on the environment. This is termed a "departure." Any deviation from the natural condition is considered an "impairment." The NPS believes the departure analysis method to be valid, and although it has not been tested in court, it has resulted in several successful water rights settlements (Gillilan and Brown, 1997).

A recent method proposed by Richter et al. (1997), the RVA or "Range of Variability Approach," focuses on the critical role of hydrologic variability in sustaining aquatic ecosystems. This method uses measured or synthesized daily streamflow data from a period during which human alteration of the hydrologic system was negligible; this streamflow record is then characterized using 32 variables defined by Richter et al. (1996). A range in variation of each of these 32 parameters is then determined and translated into management strategies. The RVA is not intended to be prescriptive; rather, it provides river managers with an interim management strategy to estimate the hydrological needs of aquatic and riparian ecosystems.

Despite the difficulty in estimating instream flows for ecological uses, the NWUIP should initiate efforts to develop such techniques on a nationwide basis. The NWUIP should collaborate with other branches of the USGS, particularly the MESC group in Fort Collins, Colorado and other agencies in the federal government.

ASSESSING THE RELEVANCE OF WATER USE INFORMATION IN DETERMINING INSTREAM WATER AVAILABILITY

In October 2000, the committee heard Dr. Gregory E. Schwarz, USGS Branch of Systems Analysis, present a holistic approach for the NWUIP. Dr. Schwarz discussed the 1978 Second National Water Assessment of the Water Resources Council, which provided an overview of the nation's 99 water resource assessment regions and of the nation as a whole. This assessment provided estimates of natural streamflow, groundwater overdraft, reservoir overdraft, and consumptive use and then used a mass balance to calculate the streamflow leaving each assessment region. Dr. Schwarz argued for such a holistic approach because it provides (1) important checks on the accuracy of the source data, (2) a framework for organizing and analyzing source data, and (3) information on where water resources are scarcest and accurate source data are most needed (G.E. Schwarz, USGS, personal communication, 2000).

Incorporation of the SPARROW Model

The approach proposed by Dr. Schwarz would incorporate the surface water quality model SPARROW (Smith et al., 1997) with water use data, including consumptive and nonconsumptive uses. SPARROW is capable of integrating watershed data over multiple spatial scales. It would be used with the EPA's River Reach File 1 (RF1) (DeWald et al., 1985), a 1:500,000-scale digital stream dataset that is attributed with stream-reach length, average stream discharge, and average flow velocity. The idea behind this approach is to determine the extent to which water use affects measured flow; land use could also be determined, thus providing a means of assessing land use–water use relationships. In theory, streamflow estimates could be made for every reach in the RF1 stream network. A streamflow model, similar in structure to the SPARROW model for nutrient loads (Preston and Brakebill, 1999), would be estimated using actual gaging station data as the dependent variable. The regression model would preserve mass balance and could be used to make predictions for every reach in the network.

One of the explanatory variables that could be used in such a regression model would be the amount of water withdrawn by use category. A coefficient would be estimated for each category. The negative value of this coefficient might be interpreted as the share of withdrawn water that is consumed by that use category. The statistical significance of these coefficients would give an indication of whether the water use data are of reasonable accuracy to statistically reflect the effect of the water use on the availability of water in the stream. The magnitude of the coefficients would determine if there are any gross biases in the water use estimates of the proportion of water that is consumed (G.E. Schwarz, USGS, personal communication, 2000).

An error analysis of the results could also be performed, whereby squared values of the estimated errors would be correlated with various terms in the model. The significance of these correlations would indicate potential sources of error in the model. Presumably, if the water use variables were correlated with the squared error, then they would represent a significant source of error in the model. There are two interpretations that could be drawn from this correlation: the error is due to systematic error in the water use estimate, and/or the error is due to random variation in the use coefficient. In either case, the magnitude of the correlation and the size of the error could be used to improve the estimate of consumptive use in a watershed. Thus, such a model would represent a holistic analysis of water availability that would parallel the approach of the Second National Water Assessment (G.E. Schwarz, USGS, personal communication, 2000).

Such an approach would not represent an alternative method for estimating water use. Rather, it would be a tool for assessing the relevance of water use information in determining water availability in the stream, which might prove useful in estimating instream flows for ecological needs.

Various sampling and other methods could be used to make independent estimates of water use in the absence of detailed water use estimates in every state. Site-specific survey information, wherever it is available, could be linked to ancillary data that are available everywhere. Predictions of water use for regions with no water use information could then be based on extrapolations of the survey data using the ancillary information. Standard errors of the estimates could also be computed. The primary challenge for this approach would be the identification of reasonable variables for the ancillary information. As has been described in earlier chapters, the accuracy of such variables varies with water use category.

If successful, the advantages of such a model would include the following: (1) model water use estimates would include a measure of error, (2) model estimates would be consistent with streamflow (and perhaps groundwater) data, improving the integrity of the estimates, (3) land use and water use relationships could be discerned, and (4) the model would represent a holistic approach to water resource assessment. Implementation of this model approach would require an inventory of the information available in each state for calculating water use, identification of the population of water users for extrapolation of survey data, and determination of the location of this population relative to the reach network for synthesis with other water data (G.E. Schwarz, USGS, personal communication, 2000).

CONCLUSIONS AND RECOMMENDATIONS

It is becoming increasingly apparent that water quality, land use, and water use cannot be considered in isolation. An approach that integrates land use, water quality, and water use would provide a great service to the nation. The USGS, through the NWUIP, could provide desperately needed leadership in this arena.

The NWUIP should be expanded to encompass water quality and the effects of land use on water use. Collaboration between the NWUIP and NAWQA programs should be strengthened.

The estimation of instream flows for ecological needs, although quite daunting, will nonetheless become increasingly important as society seeks to maintain aquatic and riparian ecosystems while also using as much of a stream's flow for nonecological needs. Most current methods for estimating instream flows are species-specific and cannot estimate instream flow requirements for multispecies assemblages or entire aquatic/riparian ecosystems. The specter of global climate change also looms large, as it may diminish flows in already overallocated streams.

The NWUIP should encourage the development of methods to estimate instream flows for ecological needs. Collaboration with organizations already engaged in this research (e.g., MESC, The Nature Conservancy) should be pursued.

Some water-quality models such as SPARROW are capable of integrating watershed data over multiple spatial scales. Such a model could be used with River Reach File 1 (RF1) to provide estimates of how water use affects streamflow, which could then be incorporated into a national water use model. This approach would not provide an alternative method for estimating streamflow, but rather would serve as a tool for assessing the relevance of water use and land use information in determining water availability in a stream. It would form part of larger holistic approach to national water resource assessment and water use.

The integration of water use data into water-quality models has the potential to greatly strengthen the present NWUIP. Steps should be taken to determine the feasibility of this approach.

9

Conclusions and Recommendations

The National Water-Use Information Program (NWUIP) is responsible for the collection and dissemination of data on the use of water resources within the United States. This is the nation's only assessment of such data. As the nation's source of unbiased science-based information on water resources, the U.S. Geological Survey (USGS) is uniquely qualified to compile and provide water use information. This information is essential to the nation to maintain a national water inventory, assure the nation's water supply, evaluate whether water supplies are sufficient for current and future needs, and evaluate the quality and quantity of water available for ecological resources.

The NWUIP is different from other water resources programs of the USGS. It is the only USGS water resources program in which the USGS does not have the principal responsibility for primary data collection. Instead, the USGS generally compiles existing data from other federal, state, and local sources. The quality of these data, therefore, is highly variable and depends upon the interests of the state or local cooperators. For this reason there are significant differences among USGS districts in data collection and/or estimation procedures, quality assessment procedures, and data distribution and accessibility. Financial support for the NWUIP largely comes from the USGS Cooperative Water (Coop) Program, whose funds are generally unavailable in states making little effort to collect water use data. This makes it difficult for USGS district offices in states not providing matching funds to compile their state's information for the five-year national water use summary reports (a key product of the NWUIP). These reports, "Estimated Water Use in the United States," have been published every five years since 1950 and are one of the most widely cited publications of the USGS.

Water use data are collected at various spatial scales: for political units such as counties and states and for water resources units such as river basins, aquifers, and hydrologic units. The merit of using political units is that ancillary information on population and economic variables is readily available for them if indirect estimates of water use are needed. The merit of using water resources units is that water use can readily be linked to water availability and to the annual water budget. Where available, site-specific water use data can be summed over any required geographic area to create aggregated water use estimates.

The USGS collects water use data for a set of water use categories, and it differentiates the use of surface water from groundwater. The water use categories include public water supply, industrial and commercial use, irrigation, livestock and domestic use, and, in some states, other water uses such as aquaculture and mining. Of these categories, the most systematic data are available for public water supply. Local water utilities maintain extensive records of water pumping and household water use required for billing. Supplemental data (including population served, source waters, and treatment processes) are available for every public water supply system through the U.S. Environmental Protection Agency (EPA) Safe Drinking Water Information System.

Industrial and commercial water use data are less systematically collected across the nation, although some states maintain comprehensive inventories of these data. Irrigation water use data are highly variable in quality, depending on reporting requirements at the level of the state, the watershed, or the irrigation project. Livestock and rural domestic water use involves small amounts of water that often fall below the trigger levels used in state water use data collection programs. No states routinely collect, or require reporting of, instream water needs or ecological water requirements, even in regions where such ecological estimates of water use are used in water resource allocation decisions.

With assistance from the NWUIP, the committee surveyed water use data collection practices for all 50 states, Puerto Rico, and the District of Columbia. This survey showed that most states collect water use data annually, if at all, but about 10 states maintain a comprehensive site-specific water use database of monthly water use data. Monthly data are valuable because they quantify the seasonal variability of water use, especially for irrigated agriculture, and they support monthly computation of the hydrologic water balance in watersheds and river basins. A more comprehensive survey and analysis of state water use data collection programs would be very valuable and would have many useful applications.

The committee's examination of NWUIP data suggests that statistical sampling and estimation methods can be applied to derive water use estimates for different spatial scales, such as states, counties, river basins, aquifers, and hydrologic units. In one case study, the committee examined data on irrigation withdrawals in Arkansas, a state with a "complete" inventory of many water use categories, to investigate the standard error in state-level estimates that would have been achieved by stratified random sampling of a small percentage of permitted withdrawals. The results of these preliminary investigations were very encouraging. They suggest real potential for the NWUIP to make greater use of statistical sampling to obtain water use estimates for states that, unlike Arkansas, do not do a complete inventory. Study of data from other states would, of course, be needed to refine appropriate sampling strategies.

In another case study, the committee analyzed the structure of the 1980–1995 state-level data from the NWUIP by multiple regression analysis in order to determine if aggregate water use could be correlated with routinely collected demographic, economic, and climatic data. Indeed, a number of potential explanatory variables for water use were identified for several major water withdrawal categories. These included water price and gross state product for public water supply, and the existence of "closed-loop" systems (i.e., cooling towers) for thermoelectric withdrawals. Significantly, the analysis also found discernable responses to climate variability in aggregate state-level data. This suggests the feasibility of normalizing water use estimates for interannual climate variability using statistical techniques.

These analyses suggest how statistical methods may yield rigorous estimates and quantitative confidence limits for aggregate water use. Where such an inventory is cost-effective, water use estimates should be based on a complete inventory of significant water use sites in a region. In practice, water use estimates compiled for each category and county in the nation will consist of a combination of direct observation, random sampling, modeling, and statistical estimation. This mixture of water use estimation techniques requires a framework that best combines data from direct inventories with estimates from statistical sampling and modeling. In this framework, the NWUIP would be supported by water use *data*, water use *estimation*, and integrative water use *science*.

The integrative science that is needed to complement the data and estimation methods described above is rarely done. Consequences of water use practices have been a focus of several new National Water-Quality Assessment (NAWQA) studies, illustrating the potential utility of integrating water quality, quantity, and use. However, water use data are usually not a major component of scientific studies conducted by the USGS. The minimal level of integration of water use data into USGS hydrologic studies is not healthy for the USGS or the nation.

The committee also found significant potential for the NWUIP to coordinate its data collection efforts more closely with activities of other agencies and institutions in order to integrate national databases in the NWUIP. The USDA maintains a five-year Census of Agriculture, which inventories irrigated agriculture using a stratified random sampling approach similar to that suggested in this report for water use. The EPA maintains a Safe Drinking Water Information System and a Permit Compliance System, both of which contain valuable inventories of water-using facilities. The Energy Information Administration maintains detailed operating information on the nation's power plants that may be useful for estimating thermoelectric water use. Site-specific water use datasets compiled at the state level also provide a rich store of information. Integrating these state and national databases will support the rigorous systematic evaluation of water use sampling and estimation methods recommended by the committee.

RECOMMENDATIONS

Many recommendations for improving the NWUIP are found within the individual chapters of this report. Of these, there are five comprehensive recommendations that will lead to a more scientifically based and better-integrated water use program. These recommendations are summarized in the following paragraphs.

Create a Water Use Science Program

The NWUIP appears to be viewed as a water use accounting program by many scientists in the USGS Water Resources Division. NWUIP needs to be elevated to being an integral part of the scientific data collection and dissemination mission of the agency. The program would significantly benefit from rigorous, hypothesis-driven research to improve our knowledge and understanding of the role of water use in hydrologic and hydrogeochemical processes and to enable the systematic development of rigorous statistical techniques for national water use estimation. Because of the significant differences in water use data collection procedures and data quality from state to state, water use accounting alone cannot provide the estimates of water use that are needed by the nation.

Rather, integrative water use science will require a more complete data paradigm, capable of tracking the sources and fate of water, as it flows between the infrastructure and natural water systems. A "link-node" representation of the movement of water from points of withdrawal, through the infrastructure water system, to points of water discharge, and back to the natural water system is one useful approach. This comprehensive and integrated approach creates the data structures and framework needed to track changes in both water quantity and quality as water moves through the landscape.

The intimate linkage between water quality and water quantity creates both opportunity and need for better connections and integration between the NWUIP, the NAWQA Program, and the USGS Biological Resources Division (BRD). Water requirements for ecological uses of instream water compete with water withdrawals from streams for other uses. In integrative water use science, both withdrawal uses and instream uses are logically parts of the total water use, within which the effects of human water uses on ecological resources can be studied. The linkage of water use infrastructure with natural water systems offers a robust framework for understanding the impacts of water use on water quality, on aquatic ecosystems, on the hydrologic cycle, and on the reliability and vulnerability of the nation's water resources.

Recommendation: The NWUIP should be elevated to *a water use science* program, emphasizing applied research and techniques development in the statistical estimation of water use and the determinants and impacts of water-using behaviors.

Improve the Water Use Database

The current water use program does not have a systematic approach to evaluating the accuracy of water use data. Although the quality of the NWUIP's historical data can be improved with consistency checks, with analysis of model residuals, and with other quality assurance techniques, the resultant data sets will not be error-free. The highest-quality water use data will come from comprehensive data collection programs that field-verify primary water use data collected at sample sites.

The committee's survey of existing water use datasets has revealed that about 20 states have comprehensive site-specific databases of annual water use, and most other states have water use data for some water use categories. In addition, the federal government systematically collects information about water use facilities just as it collects national economic, population, and other resource data relevant to water use estimation. These data can be integrated and synthesized at the national level to systematically support more rigorous water use science in the United States. The convergence of consistent national data sets and widely available standard technologies in geographic information systems creates a timely opportunity for USGS to improve the acquisition and management of data used to estimate national water use.

Recommendation: To better support water use science, the USGS should build on existing data collection efforts to systematically integrate datasets, including those maintained by other federal and state agencies, within the data collection and water use estimation activities of the NWIUP.

Improve Water Use Estimation with Improved Statistical Procedures

National summaries of water use are produced every five years by the USGS using an inventory approach in which an estimate is made of the total water use in each of a set of water use categories for each county in a state. The state totals are then summed to create national totals. To the extent possible, county and state estimates are based on information from surveys of water users, supplemented by indirect estimates where surveyed water use data are not available. Statistical investigations carried out by the committee suggest the potential for improvement that could be realized from conducting the estimation of state and national water use through a rigorous framework of statistical sampling and estimation. Research on statistical estimation of water use should include a systematic evaluation of the errors of reported and measured major water uses. This would strongly complement the USGS's expertise and activities in other national statistical synthesis efforts such as the NAWQA Program.

Recommendation: The USGS should systematically compare water use estimation methods to identify the techniques best suited to the requirements and limitations of the NWUIP. One goal of this comparison should be to determine the standard error for every water use estimate.

Integrate Water Use Information into USGS Water Resources Research Programs

Human use of water is highly dependent upon the availability of water, which is susceptible to climate change. Human water use is one of the main mechanisms changing the spatial and temporal distribution of the natural water resource system. Water use should be viewed as a component of the hydrologic budget for watersheds, aquifers, and river basins. Water use estimates should be used to assess the impact of human activity on the quantity and quality of water within the hydrologic systems of interest. Water use data can be used to evaluate water use trends and to estimate future water needs in river basins or aquifers where water resources might be limited.

An integrative water use science program that investigates spatial and temporal patterns of water use, impacts of water use on aquatic ecosystems and the hydrologic cycle, and the sustainability, reliability, and vulnerability of water resources cannot stand alone. It needs to leverage the resources of other USGS programs. An example would be an investigation of the feasibility of modeling approaches (such as adaptations of the water-quality model SPARROW) to assess the relevance of water and land use information in determining water availability. Another example is developing a framework for incorporating estimates of instream flow used for ecological purposes. Collaboration with the Biological Resources Division and with state and federal resource management agencies will be necessary to develop such approaches.

Recommendation: The USGS should focus on the scientific integration of water use, water flow, and water quality in order to expand knowledge and generate policy-relevant information about human impacts on both water and ecological resources.

Redefine the Goals of the Water Use Program and Improve the Status of the Program Within the USGS

In this evaluation of the NWUIP, the committee found a program that is a unique source of critically needed water use data for the nation and for other water resources programs within the USGS. This report envisions a substantial transformation of the NWUIP. The committee does not see the NWUIP as simply a database management program focused on county-level categorical water use. Rather, the committee finds a natural role for the NWUIP to complement and become actively integrated with the Survey's other efforts to provide unbiased science-based information about the nation's water resources. This focus requires a core national effort to supplement the activities presently carried out by the water use programs in the states.

The USGS must find a stable source of dedicated funds to support a continuous, sustained effort to collect, analyze, evaluate, and disseminate water use data. The sole reliance on Coop Program funds has resulted in a national program that is based on data collection procedures with variable regional coverage and weak control of the quality of the data. This is inconsistent with the Survey's role as the nation's unbiased source of water resources information.

Recommendation: The USGS should seek support from Congress for dedicated funding of a national component of the recommended water use science program. This funding would supplement the existing funding in the Coop Program.

References

Allen, R. G., W. O. Pruitt, J. A. Businger, L. J Fritschen, M. E. Jensen, and F. H. Quinn. 1996. Evaporation and Transpiration. Chapter 4, p. 125-252 in: Wootton et al. (Ed.), American Society of Civil Engineers Handbook of Hydrology. ASCE: New York, NY.

Alley, W. M. 1984. The Palmer Drought Severity Index: Limitations and Assumptions. Journal of Climate and Applied Meteorology 23:1100-1109.

Anderson, M., and R. Magleby. 1997. Agricultural Resources and Environmental Indicators, 1996-1997. U.S. Department of Agriculture Economic Research Service Agricultural Handbook No. 712. Washington, D.C.

Auble, G. T., J. M. Friedman, and M. L. Scott. 1994. Relating riparian vegetation to present and future streamflows. Ecological Applications 4:544-554.

Billings, R. B., and C. V. Jones. 1996. Forecasting Urban Water Demand. American Water Works Association, Denver, CO.

Biviano, M. B., D. E. Sullivan, and L. A. Wagner. 1999. Total Materials Consumption: An Estimation Methodology and Example Using Lead—A Materials Flow Analysis. USGS Circular 1183. U.S. Geological Survey: Reston, Va.

Boland, J. J., B. Dziegielewski, D. D. Baumann, and E .M. Opitz. 1984. Influence of Price and Rate Structures on Municipal and Industrial Water Use. IWR Report 84-C-2. U. S. Army Corps of Engineeers Institute for Water Resources, Fort Belvoir, Virginia.

Bovee, K. D., and R. T. Milhous. 1978. Hydraulic simulation in instream flow studies: Theory and techniques. Instream Flow Information Paper 5. U.S. Fish and Wildlife Service Biological Report FWS/OBS-78/33. 130 p.

Buchmiller, R., M. Focazio, L. Franke, W. Freeman, S, Gain, E. Josberger, and C. Tate. 2000. A Vision for the U.S. Geological Survey National Water-Use Information Program. Unpublished report prepared by the internal USGS planning team on future of the National Water-Use Assessment Program. U.S. Geological Survey: Reston, Va.

Castleberry, D. T., J. J. Cech, D. C. Erman, D. Hankin, M. Healey, G. M. Kondolf, M. Mangel, M. Mohr, P. B. Moyle, J. Nielsen, T. P. Speed, and J. G. Williams. 1996. Uncertainty and instream flow standards. Fisheries 21:20-21.

Christensen, V. G., X. Jian, and A. C. Ziegler. 2000. Regression Analysis and Real-Time Water-Quality Monitoring to Estimate Constituent Concentrations, Loads, and Yields in the Arkansas River, South-Central Kansas, 1995-99. USGS Water-Resources Investigations Report 00-4126. Lawrence, Kansas.

Cochran, W. G. 1977. Sampling Techniques. Third Edition. John Wiley and Sons: New York.

Cohn, T. A., L. L. Delong, E. J. Gilroy, R. M. Hirsch, and D. K. Wells. 1989. Estimating Constituent Loads. Water Resources Research, 25(5), 937-942.

Dash, R.G., B. M. Troutman, and P. Edelmann. 1999. Comparison of Two Approaches for Determining Ground-Water Discharge and Pumpage in the Lower Arkansas River Basin, Colorado, 1997–98. USGS Water-Resources Investigations Report 99–4221. U.S. Geological Survey Denver, CO.

Davis, W. Y., D. M. Rodrigo, E. M. Opitz, B. Dziegielewski, D. D. Baumann, and J. J. Boland. 1991. IWR-MAIN water use forecasting system, version 5.1—users manual and system description, consultant report: Carbondale, Ill., U.S. Army Corps of Engineers and Planning and Management Consultants, 307 p.

Department of Water Resources. 1998. California Water Plan Update. State of California, Department of Water Resources Bulletin 160-98: Volume 1, Chapter 4.

DeWald, T., R. Horn, R. Greenspun, P. Taylor, L. Manning, and A. Montalbano. 1985. STORET Reach Retrieval Documentation. U.S. Environmental Protection Agency.

Dziegielewski, B., S. Sharma, T. Bik, X. Yang, and H. Margono. 2002a. Analysis of Water Use Trends in the United States: 1950-1995. Report for USGS Water Resources Program. Southern Illinois University. Carbondale, Illinois (Forthcoming).

Dziegielewski, B., S. Sharma, T. Bik, X. Yang, H. Margono, and R. Sa. 2002b. Predictive Models of Water Use: Analytical Bibliography. Southern Illinois University. Carbondale, Illinois (Forthcoming).

Dziegielewski, B., E. M. Opitz, and D. Maidment. 1996. Chapter 23. Water Demand Analysis. p.23.1-23.62, In: (L. Mays, Ed.) Water Resources Handbook, McGraw-Hill.

Garcia-Acevedo, M. R. 2001. The confluence of water, pattern of settlement, and construction in the Imperial and Mexicali Valleys (1900-1999). Pp. 67-88 in Reflections on Water: New Approaches to Transboundary Conflicts and Cooperation (J. Blatter and H. Ingram, eds.), Cambridge, MA, MIT Press, 358 p.

Getches, D. H., L. J. MacDonnell, and T. A. Rice. 1991. Controlling Water Use: The Unfinished Business of Water Quality Protection. Natural Resources Center, University of Colorado School of Law.

Gillilan, D. M., and T. C. Brown. 1997. Instream Flow Protection. Island Press, 417 p.

Gilliom, R. J., W. M. Alley, and M. E. Gurtz. 1995. Design of the National Water Quality Assessment Program—Occurrence and Distribution of Water-Quality Conditions. U.S. Geological Survey Circular 1112. 33 p.

Gisser, M. 1970. Linear programming models for estimating the agricultural demand function for imported water in the Pecos River basin. Water Resources Research 6(4):1025-1032.

Guttman, N. B., J. R. Wallis, and J. R. M. Hosking. 1992. Spatial Comparability of the Palmer Drought Severity Index. Water Resources Bulletin 28:1111-1119.

Guyton, W. F. 1950. Estimated Use of Groundwater in the United States, 1945. Unpublished report. U.S. Geological Survey.

Hanemann, W. M. 1998. Determinants of Urban Water Use. In: Urban Water Demand Planning and Management. D. D. Baumann, J. J. Boland, and W. M. Hanemann (Eds.). McGraw-Hill, Inc.

Helsel, D. R., and R. M. Hirsch. 1992. Statistical Methods in Water Resources. Elsevier: Amsterdam.

Hirsch, R. M., W. M. Alley, and W. G. Wilber. 1988. Concepts for a National Water-Quality Assessment Program. U.S. Geological Survey Circular 1021. 42 p.

Horne, M. A. 2001. U.S. Geological Survey, Personal Communication.

Howitt, R. H., J. R. Lund, W. Kirby, M. W. Jenkins, A. J. Draper, P. M. Grimes, K. B. Ward, M. D.
    Davis, B. D. Newlin, B. J. Van Lienden, J. L. Cordua, and S. M. Msangi. 1999. Integrated
    Economic-Engineering Analysis of California's Future Water Supply. Report for the State of
    California Resources Agency. Department of Agriculture and Resource Economics. Davis,
    CA.
Jensen, M. E., R. D. Burman, and R. G. Allen (ed.). 1990. Evapotranspiration and Irrigation Water
    Requirements. American Society of Civil Engineers, Engineering Practices Manual No. 70.
    332 p. ASCE: New York, NY.
Karpisack, M., and M. H. Marion. 1994. Evaporative Cooler Water Use. Publication 191045.
    University of Arizona Cooperative Extension, Tucson, AZ.
Kliot, N., D. Shmueli, and U. Shamir. 1997. Institutional Frameworks for the Management of
    Transboundary Water Resources. Volume I: Institutional Frameworks Reflected in Thirteen
    River Basins. Haifa, Israel, Water Research Institute, Technion-Israel Institute of Technology,
    417 p.
Koeppel, G. T. 2000. Water for Gothan: A History. Princeton University Press: Princeton, NJ.
Lamb, B., and H. Doerksen. 1990. Instream water use in the United States-water laws and methods
    for determining flow requirements. Pp. 109-116 in National Water Summary 1987—Hydro-
    logic Events and Water Supply and Use. U.S. Geological Survey Water-Supply Paper 2350.
    553 p.
Leahy, P. P., J. S. Rosenshein, and D. S. Knopman. 1990. Implementation Plan for the National
    Water-Quality Assessment Program. U.S. Geological Survey Open-File Report 90-174. 10 p.
Lumia, D. S. 2000. History of the National Water-Use Information Program. Unpublished report.
    USGS: Troy, NY. 5. p.
MacDonnell, L. J. 1999. From Reclamation to Sustainability: Water, Agriculture and the Environ-
    ment in the American West. Niwot, CO: University Press of Colorado, 385 p.
MacKichan, K. A. 1951. Estimated Water Use in the United States in 1950. U.S. Geological
    Survey Circular 115. 13 p.
MacKichan, K. A. 1957. Estimated Water Use in the United States in 1955. U.S. Geological
    Survey Circular 398. 18 p.
MacKichan, K. A., and J. D. Kammerer. 1961. Estimated Use of Water in the United States in 1960.
    U.S. Geological Survey Circular 456. 26 p.
Meyer, J. L., M. J. Sale, P. J. Mulholland, and N. L. Poff. 1999. Impacts of climate change on
    aquatic ecosystem functioning and health. Journal of the American Water Resources Associa-
    tion 35:1373-1386.
Minnick, R., and J. S. Parrish. 1994. Public-land surveys. Pp. 739-744 in The Surveying Hand-
    book, R. C. Brinker and R. Minnick (eds.). Chapman and Hall: New York.
Moore, M., N. Gollehon, and M. Carey. 1994a. Alternative models of input allocation in multicrop
    system: Irrigation water in central plains. Agricultural Economics 11:143-158.
Moore, M., N. Gollehon, and M. Carey. 1994b. Multicrop production decisions in western irrigated
    agriculture: The role of water price. American Journal of Agricultural Economics 76:859-874.
Morehart, M., N. Gollehon, R. Dismukes, V. Breneman, and R. Heimlich. 1999. Economic Assess-
    ment of the 1999 drought: Agricultural Impacts are Severe Locally, but Limited Nationally.
    U.S. Department of Agriculture, Economic Research Service, Agriculture Information Bulletin
    No. 755. Washington, D.C.
Mullusky, M. G., S. S. Schwartz, and R. C. Steiner. 1995. Water Demand Forecast and Resource
    Availability Analysis for the Washington Metropolitan Area. Interstate Commission on the
    Potomac River Basin Technical Report 95-6. Rockville, MD.
Murray, C. R. 1968. Estimated Use of Water in the United States in 1965. U.S. Geological Survey
    Circular 556. 53 p.
Murray, C. R., and E. B. Reeves. 1972. Estimated Use of Water in the United States in 1970: U.S.
    Geological Survey Circular 676. 37 p.

Murray, C. R., and E. B. Reeves. 1977. Estimated Use of Water in the United States in 1975. U.S. Geological Survey Circular 765. 37 p.

Muth, R. T., L. W. Crist, K. E. LaGory, J. W. Hayse, K. R. Bestgen, T. P. Ryan, J. K. Lyons, and R. A. Valdez. 2000. Flow and temperature recommendations for endangered fishes in the Green River downstream of Flaming Gorge Dam. Final Report. Upper Colorado River Endangered Fish Recovery Program Project FG-53.

National Research Council. 1989. Irrigation-Induced Water Quality Problems. National Academy Press: Washington, D.C.

National Research Council. 1992. Restoration of Aquatic Ecosystems: Science: Technology and Public Policy. National Academy Press: Washington, D.C.

National Research Council. 1999. Nature's Numbers: Expanding the National Economic Accounts to Include the Environment. National Academy Press: Washington, D.C.

National Research Council. 2001. Aquifer Storage and Recovery in the Comprehensive Everglades Restoration Plan: A Critique of the Pilot Projects and Related Plans for ASR in the Lake Okeechobee and Western Hillsboro Areas. National Academy Press: Washington, D.C.

Newlin, B. D., M. W. Jenkins, J. R. Lund, and R. E. Howitt. 2001. Southern California water markets: Potential and limitations. Journal of Water Resources Planning and Management.

Nielsen, D. C. 1995. Conservation Tillage Fact Sheet #2-95. Water Use/Yield Relationships for Central Great Plains Corps. Natural Resources Conservation Service, Agriculture Research Center, and Colorado Conservation Tillage Association, Akron.

Owen-Joyce, S. J., and L. H. Raymond. 1996. An accounting system for water and consumptive use along the Colorado River, Hoover Dam to Mexico. U.S. Geological Survey Water-Supply Paper 2407, 94 p.

Picton, W. L. 1952. The national picture. Illinois State Water Survey Bulletin 41:127-130.

Pierce, R. R. 1993. The National Water-Use Information Program: State of the Program 1993. Unpublished USGS NWUIP report. 17 p.

Poff, N. L., J. D. Allan, M. B. Bain, J. R. Karr, K. L. Prestegaard, B. Richter, R. Sparks, and J. Stromberg. 1997. The natural flow regime: A new paradigm for riverine conservation and restoration. BioScience 47:769-784.

Preston, S. D., and J. W. Brakebill. 1999. Application of Spatially Referenced Regression Modeling for Total Nitrogen Loading in the Chesapeake Bay Watershed. U.S. Geological Survey Water-Resources Investigations report 99-4054. 12 p.

Richter, B. D., J. V. Baumgartner, J. Powell, and D. P. Braun. 1996. A method for assessing hydrologic alteration within ecosystems. Conservation Biology 10:163-1174.

Richter, B. D., J. V. Baumgartner, R. Wigington, and D. P. Braun. 1997. How much water does a river need? Freshwater Biology 37:231-249.

Rood, S. B., and J. M. Mahoney. 1990. Collapse of riparian poplar forests downstream from dams in western prairies: probable causes and prospects for mitigation. Environmental Management 14:451-464.

Schaible, G., N. Gollehon, M. Kramer, M. Aillery, and M. Moore. 1995. Economic Analysis of Selected Water Policy Options for the Pacific Northwest. Agricultural Economic Report-720.

Schwartz, S. S., and D. Q. Naiman. 1999. Bias and variance of planning level estimates of contaminant loads. Water Resources Research 35:3475-3487.

Smith, R. A., G. E. Schwarz, and R. B. Alexander. 1997. Regional interpretation of water-quality monitoring data. Water Resources Research 33:2781-2798.

Snavely, D. S. 1986. Water-Use Data-collection Programs and Regional Data Base of the Great lakes-St. Lawrence River Basin States and Provinces. U.S. Geological Survey Open-File Report 86-546. 204 p.

Solley, W. B., E. B. Chase, and W. B. Mann, IV. 1983. Estimated use of water in the United States in 1980: U.S. Geological Survey Circular 1001. 56 p.

Solley, W. B., C. F. Merk, and R. R. Pierce. 1988. Estimated Use of Water in the United States in 1985. U.S. Geological Survey Circular 1004. 82 p.

Solley, W. B., R. R. Pierce, and H. A. Perlman. 1993. Estimated Use of Water in the United States in 1990. U.S. Geological Survey Circular 1081. 76 p

Solley, W. B., R. R. Pierce, and H. A. Perlman. 1998. Estimated Use of Water in the United States in 1995. U.S. Geological Survey Circular 1200. 71 p.

Sparks R. E. 1992. Risks of altering the hydrologic regime of large rivers. Pages 119-152 in J. Cairns, B. R. Niederlehner, and D. R. Orvos (eds.). Predicting ecosystem risk. Vol XX: Advances in modern environmental toxicology. Princeton, NJ: Princeton Scientific Publishing Co.

Stalnaker, C., B. Lamb, J. Henriksen, K. Bovee, and J. Bartholow. 1995. The Instream Flow Incremental Methodology-A Primer for IFIM. U.S. Department of the Interior, National Biological Service, Technical Report 29. 45 p.

Tarlock, A. D. 1999. The creation of new risk based water sharing entitlement regimes: The case of the Truckee-Carson settlement. Ecology Law Quarterly 25:674-691.

Thompson, S. A. 1999. Water Use: Management and Planning in the United States. Academic Press: San Diego, Calif.

U.S. Bureau of Reclamation. 2000. River and Reservoir Operations Simulation of the Snake River: Application of MODSIM to the Snake River Basin. U.S. Department of the Interior, Bureau of Reclamation, Pacific Northwest Region.

U.S. Department of Agriculture National Agricultural Statistics Service. 1999a. 1997 Census of Agriculture. United States Summary and State Data. Geographic Area Series no. AC97-A-51. Online. Available: U.S. Department of Agriculture. http://www.nass.usda.gov/census/. Accessed July 20, 2001.

U.S. Department of Agriculture National Agricultural Statistics Service. 1999b. 1998 Farm & Ranch Irrigation Survey. Volume 3, Part 1 Special Studies of 1997 Census of Agriculture, AC97-SP-1. U.S. Government Printing Office, Washington, DC. Also, online at http://www.nass.usda.gov/census/census97/fris/fris.htm. Accessed January 7, 2002.

U.S. Environmental Protection Agency. 2000. Water Quality Conditions in the United States: A Profile from the 1998 National Water Quality Inventory to Congress. Report EPA841-F-00-006. U.S. Government Printing Office: Washington, D.C. 135 p.

U.S. Geological Survey. 1981. National Water-Use Information Program. U.S. Geological Survey Fact Sheet. Reston, VA: U.S. Geological Survey.

U.S. Geological Survey. 1999. The Quality of Our Nation's Waters –Nutrients and Pesticides. U.S. Geological Survey Circular 1225. U.S. Geological Survey: Reston, Va. 82 p.

U.S. Geological Survey. 2000. National Handbook of Recommended Methods for Water Data Acquisition. Chapter 11—Water Use. Prepared under the sponsorship of the U.S. Geological Survey Office of Water Data Coordination. Online. U.S. Geological Survey. Available: http://water.usgs.gov/pubs/chapter11/. Accessed July 20, 2001.

Vörösmarty, C. J., P Green, J. Salisbury, and R. B. Lammers. 2000. Global Water Resources: Vulnerability from Climate Change and Population Growth. Science 289:284-288.

Wernik, L. K., and J. H. Ausebel. 1995. National materials flows and the environment. Annu. Rev. Energy Environ. 20:463-492.

Whitcomb, J. B. 1988. A Daily Municipal Water-Use Model: Case Study Comparing West Los Angeles, California and Fairfax County, Virginia. Dissertation. The Johns Hopkins University, Baltimore, MD.

Williams, J. G. 1996. Lost in space: minimum confidence intervals for idealized PHABSIM studies. Transactions of the American Fisheries Society 125:458-465.

Woodwell, J., and P. Desjardin. 1995. Analysis of Household and Jobs Per Acre, 1990 and 2020 – From COG's Round 5.2 Cooperative Forecasts. Department of Human Services, Planning and Public Savety, Metropolian Washington Council of Governments, Washington, D.C.

Acronyms and Abbreviations

ABF	aquatic base flow
APE	absolute percent error
ASWCC	Arkansas Soil and Water Conservation Commission
AWUDS	Aggregated Water-Use Data System
BEA	Bureau of Economic Analysis (U.S. Department of Commerce)
BLM	Bureau of Land Management
BRD	Biological Resources Division (U.S. Geological Survey)
cfs	cubic feet per second
Coop	USGS Cooperative Water Program
Corps	U.S. Army Corps of Engineers
EIA	Energy Information Administration (U.S. Department of Energy)
EMAP	Environmental Monitoring and Assessment Program (U.S. Environmental Protection Agency)
EPA	U.S. Environmental Protection Agency
ERS	Economic Research Service (U.S. Department of Agriculture)
FGDC	Federal Geographic Data Committee
FRIS	Farm and Ranch Irrigation Survey (U.S. Department of Agriculture)
gal	gallons
gpcd	gallons per capita per day
gpm	gallons per minute

GPS	global positioning system
GIS	geographic information system
HQI	habitat quality index
IFIM	instream flow incremental methodology
I-O	input-output [tables]
LHS	left-hand side (i.e., the dependent variable)
MESC	Midcontinent Ecological Science Center (U.S. Geological Survey)
MG	million gallons
MGD	million gallons per day
MSE	mean squared error
MWA	Massachusetts Water Authority
NAICS	North American Industry Classification System
NASQAN	National Stream Quality Accounting Network Program (U.S. Geological Survey)
NASS	National Agricultural Statistics Service (U.S. Department of Agriculture)
NAWQA	National Water-Quality Assessment Program (U.S. Geological Survey)
NEWUDS	New England Water-Use Data System
NEXRAD	NEXt Generation Weather RADar
NHD	National Hydrography Dataset (U.S. Geological Survey/U.S. Environmental Protection Agency)
NOAA	National Oceanic and Atmospheric Administration
NPDES	National Pollutant Discharge Elimination System (U.S. Environmental Protection Agency)
NPS	National Park Service
NRC	National Research Council
NRCS	Natural Resources Conservation Service (U.S. Department of Agriculture)
NSIP	National Streamflow Information Program (U.S. Geological Survey)
NWIS	National Water Information System (U.S. Geological Survey)
NWUDS	National Water-Use Data System (U.S. Geological Survey)
NWUIP	National Water-Use Information Program (U.S. Geological Survey)
OLS	ordinary least squares
PDSI	Palmer Drought Severity Index
PLSS	Public Land Survey System

ppm   parts per million

RF1   Reach File 1 (U.S. Environmental Protection Agency)
RHS   right-hand side (i.e., the independent variables)
RVA   range of variability approach

SDWIS   Safe Drinking Water Information System (U.S. Environmental
                 Protection Agency)
SIC   Standard Industrial Classification [codes]
SPARROW   SPAtially Referenced Regressions On Watershed attributes (U.S.
                 Geological Survey)
SRS   stratified random sampling
SU   study unit (U.S. Geological Survey/National Water Quality Assess-
                 ment Program)
SWUDS   site-specific water-use data system

TDS   total dissolved solids
TMDL   total maximum daily load
USACE   U.S. Army Corps of Engineers
USBR   U.S. Bureau of Reclamation
USDA   U.S. Department of Agriculture
USGS   U.S. Geological Survey

VOC   volatile organic compound

WRD   Water Resources Division (U.S. Geological Survey)
WSTB   Water Science and Technology Board (National Research Council)
WUDBS   Water-Use Data Base System (U.S. Geological Survey, Arkansas
                 District)

Appendix A

Narrative Description of
State Water Use Data Collection Programs

At the request of the committee, USGS water use specialists, led by the four regional water use coordinators (Deborah Lumia, Joan Kenny, Molly Maupin, and Susan Hutson) undertook a survey of the current condition of water use data collection in all 50 states, the District of Columbia, and Puerto Rico. Its purpose was to answer the question, "What kinds of data are collected and stored by each state, and how often?" A series of questions were addressed to the water use specialist in each state (see Chapter 2 for a list of the survey questions), and the responses were tabulated. This appendix is a narrative describing the results of the survey in each state. The overall results of the survey are summarized in Chapter 2.

Alabama has a permitting program for all categories of water use. Data are reported to the state annually for all public water supplies and for other all other water users whose withdrawal exceeds 100,000 gallon per day. Laws are applicable statewide with no difference between surface and groundwater. The state maintains a water use database, which is updated annually. Latitude and longitude are not recorded for wells or surface water intakes.

Alaska has a water use permitting program for all categories of water use. Monthly withdrawal amounts are requested until the facility is issued a certificate to use the water, then the facility reports only if the certificate requires reporting. Most users continue to report on a voluntary basis. Permits are required for usage in excess of 1,500 gallon per day for public water supply and domestic water use and for usage in excess of 10 acre-feet per month for all other users. Laws are applicable statewide with no difference between surface and groundwater. The state maintains a water use database, which is updated annually. Latitude and longitude are not recorded for wells or surface water intakes. Sometimes the reported water use data are checked against the permitted amounts.

Arizona requires water permits, collects water use data, and maintains a database of annual water use for all users whose groundwater withdrawal rate exceeds 35 gallon per minute in five "active management areas" (Phoenix, Tucson, Pinal, Prescott, Santa Cruz). Groundwater withdrawals are governed by a 1980 state law. Surface water has been adjudicated statewide according to the provisions of a 1919 law using the "first in time is first in right" principle. Surface water data are available from a variety of sources such as Central Arizona Project, Gila Water Commissioner, and irrigation districts. Latitude and longitude of wells and surface water intakes are not reported, but the township, range, and quarter section of these points are recorded. Data checking is done using electric power consumption data and information from satellite photos for monitoring irrigated area.

Arkansas requires water permits and maintains a database of monthly water use reported annually for all surface water uses of more than 1 acre-foot per year and all groundwater wells having a potential flow of 50,000 gallon per day or more. Latitude and longitude are required for all wells and surface water intake points, as well as township, range, section, quarter section, and quarter-quarter section for all of these points. The database is maintained by the USGS, and water use data are sometimes checked against power consumption data. Trends through time are checked for the period of the data, from 1985 to present (see Chapter 3 for more information on the Arkansas water use database).

California has the authority through the California State Water Resources Control Board to permit water use. The state requires annual reporting of surface water withdrawals for permitted water rights for the first 10 years of the permit to substantiate water use. Once the permitted withdrawal is well documented, the permit holder is issued a "license" to withdraw from surface water sources and is required to submit a "report of licensee" once every three years that documents monthly withdrawals over the last three-year period. Water rights acquired prior to 1914 and riparian water right holders are requested to report but are not required to do so. Total groundwater extractions in Los Angeles, Riverside, San Bernardino, and Ventura Counties that are in excess of 25 acre-feet per calendar year (or greater than 10 acre-feet per year from any single withdrawal point) are required to be filed with the State Water Resources Control Board within the first six months of the succeeding year. Surface water withdrawal of any amount must be reported except for the exempted water rights mentioned above. These regulations apply statewide for surface water but only to the four counties mentioned in state water code for groundwater. Many other agencies in California besides the State Water Resources Control Board collect water use information for their own purposes. Other groundwater basins have been adjudicated throughout California, and regional "water masters" have been designated by the courts to maintain groundwater extraction data. California's multiagency water management organizational structure, the number of its water users, and its complex water-transfer infrastructure combine to create a complex water use information environment.

Colorado has the legal authority to permit withdrawals, and it may require reporting of diversions of any magnitude at any location where a surface water or groundwater withdrawal occurs. Records are maintained for all major surface water deliveries. Groundwater withdrawal data are generally not required, except in the Arkansas River basin where monthly withdrawals are required for all wells pumping greater than 50 gallons per minute. Water permits are obtained statewide in the same manner for surface water as for groundwater. The location of any point of diversion may be requested by the state engineer; however, latitude and longitude generally are not required. Township, range, section, and quarter-quarter-quarter section are requested for wells. The state maintains a water use database that consists of withdrawal information originally collected at specific points of diversion, then aggregated by water district and forwarded to the state by the seven division offices. The state makes periodic site measurements and checks on site data recorders. Power companies provide electric consumption data for well pumpage calculations.

Connecticut has the legal authority to permit or register withdrawals. Public water use is reported to the state for users using greater than 50,000 gallons per day, with data being recorded monthly, quarterly, or annually depending on the user. Laws are applicable statewide and are the same for surface and groundwater. The state maintains a water use database for public water supply but not for other water use categories.

Delaware has a water use permitting program, requires the collection of data, and maintains a water use database. Monthly water use data are reported annually for any facility drawing 50,000 gallons per day or more, for all water use categories except for domestic and livestock. Both surface and groundwater use data are compiled on a similar basis. Locations of wells and water use intakes are plotted on maps. Data consistency is checked by comparison with the previous year's data at the same withdrawal site.

The **District of Columbia** has a permitting program for well construction; however, it has no authority for collecting withdrawal (or any other) data once the well is permitted. Little groundwater is pumped in the District except for dewatering or for cleanup of contaminated sites. Public water supply is managed by the U.S. Army Corps of Engineers Washington Aqueduct operations. This water comes entirely from two surface water intakes on the Potomac River located in Maryland at Little Falls and Great Falls. Monthly withdrawal information from these intakes is reported to the Maryland Department of the Environment in accordance with Maryland laws (see below).

Florida has a water use permitting program managed by five water management districts that collectively cover the state. Rules regarding trigger levels for requiring permits and the degree of reporting of water use data vary from one district to another, with the rules being more stringent in critical water use areas. In general, permits are required for all users having the capacity to use 1 million gallons per day and for all wells greater than six inches in diameter. Water use data are reported monthly, quarterly, or annually, depending on the water management district, with the exception of agricultural water use, which is collected only in some areas of the state. A database of latitude and longitude of public water supply users has been compiled, but similar location data are not generally available for other water use categories. Some data checking is done by comparison with the past year's water withdrawals.

Georgia has a water use permitting program, requires the collection of data, and maintains separate databases for surface water and groundwater use. Data are compiled annually for all users withdrawing at least 100,000 gallons per day for public, industrial, commercial, and power water use, but not for irrigation, livestock, and domestic use. Laws apply statewide in the same manner for surface and groundwater.

Hawaii has a water use permitting program and requires the collection of monthly water use data in all water use categories for uses exceeding 1,000 gallons per day. Laws apply statewide in the same manner for surface and groundwater. The database is updated monthly to quarterly, depending on the receipt of water withdrawal data. The latitude and longitude of wells are recorded; this information will be required for surface water intakes in the future.

Idaho has the legal authority to permit or register withdrawals, data are reported to the state, and a water use database is being constructed. Data reporting is required of any single- or multiple-user water system having an instantaneous diversion rate of at least 0.24 cubic feet per second (108 gallons per minute), or which is irrigating more than five acres in area. Water use regulation is mandatory for all diversions within the Eastern Snake Plain aquifer boundaries and for some diversions outside this area. The law applies similarly to surface and groundwater. The database will be updated annually. Latitude and longitude locations of wells are required and are verified using global positioning system technology. Location coordinates for surface water intakes will be required in the future. Withdrawal data are checked against water right permit limits.

Illinois does not have legal authority to permit or register withdrawals, but it does have an Illinois Water Inventory program administered by the Illinois State Water Survey. This inventory covers public water supply wells and surface water intakes, high-capacity (more than 70 gallon per minute) private wells, and surface water intakes for industry, commercial establishments, and fish and wildlife management areas. Annual water use data have been surveyed since 1978 and are stored in a Public Industrial-Commercial Survey database using township-range-section location coordinates. Agricultural water use is not systematically surveyed, except in some project areas. A separate private well database for wells with a capacity of less than 70 gallons per minute has recently been established.

Indiana has the legal authority to register water withdrawals and collects monthly data on all water use categories for any facility capable of withdrawing 100,000 gallons per day. Laws are applicable statewide and are the same for surface and groundwater. The water use database is updated annually. Locations of wells and surface intakes reported in Universal Transverse Mercator coordinates are required. The state tracks changes in withdrawals by comparing the current year's withdrawal to the previous year's withdrawal at the same site.

Iowa has a water use permit program, except for agricultural or irrigation water withdrawals from the Mississippi and Missouri Rivers, which do not require a permit. The state requires annual reporting of water use for all water uses exceeding 25,000 gallons per day. There are special threshold provisions for withdrawals from the Dakota and Jordan sandstone aquifers. Locations of wells and surface water intakes are specified by township, range, and section, but the latitude and longitude of these points are not reported. Some data checking is done against permitted usage rate and usage reports from water suppliers. The state maintains a database, which is updated annually.

Kansas has a water use permit program and collects water use data annually from all permitted water users. Annual reports for public water supply and industrial uses include monthly data. There is no lower trigger level for requiring water use reporting; all permitted users except those with domestic water rights must file an annual water use report or be fined. Permits are required for livestock water use at operations having 1,000 head or more of cattle or using at least 15 acre-feet per year for other kinds of livestock. Permits are not required for domestic use, or for public water suppliers with fewer than 10 connections. Laws are applicable statewide and are the same for surface and groundwater. Some areas of the state are closed to new appropriations. Latitude and longitude of wells and surface water intakes are assigned by the state and are verified using global positioning system technology. The water use database is updated on a continuing basis as reports are received. Annual publications of irrigation and public water use data contain five-year averages for irrigation application rates, per capita use, and unaccounted-for water (Kansas Water Office and Kansas State Department of Agriculture, 2001; Kansas Water Office and USGS, 2001).

Kentucky has a water-withdrawal permitting program and collects water withdrawal data for public water supply, industrial, commercial, and mining water use, but not for thermoelectric, livestock, domestic, or aquaculture water use. The average daily water withdrawal is reported twice a year for uses exceeding 10,000 gallons per day. Laws are applicable statewide and are the same for surface and groundwater. Latitude and longitude coordinates of wells and surface water intakes are requested but are not often supplied. Public water supply data are sometimes checked against drinking water monthly operating reports.

Louisiana has the legal authority to register and collect water use information from wells. The state collects water use data for both surface and groundwater withdrawals, although no statute exists to cover surface water information. Facilities using more than 1 million gallons per day report withdrawal information quarterly; all other facilities receive a questionnaire every five years. Aggregate information also is collected at five-year intervals. The reporting program is statewide in coverage for both surface and groundwater. Monthly water use data are collected in the Baton Rouge area by the Capital Area Ground Water Commission. The latitude and longitude location of the measuring point are provided when wells are registered. The state has an ongoing program to collect latitude/longitude information for other facilities. The state and USGS maintain a database that is updated completely every five years. Data from the major facilities (greater than 1 million gallons per day) are updated quarterly. Data are checked by comparison with previous years' values, with typical use by similar facilities, and with data from other agencies or programs such as the Louisiana Department of Health and Hospitals, National Pollutant Discharge Elimination System permits, Louisiana Department of Agriculture Extension Service, and the USDA National Agricultural Statistics Service.

Maine does not have the legal authority to permit or register water withdrawals. Public water use data are reported to the state. Public water supply data are recorded monthly and are reported annually for surface water and groundwater withdrawals for most public water utilities.

Maryland has a water use permit program for public water supply, industrial, commercial, irrigation, and power water uses, but not for domestic and livestock water use. The state maintains a water use database for which twice a year, the six-month total of water use is reported to the state. Permit holders using more than 10,000 gallons per day are required to submit reports. Laws are applicable statewide and are the same for surface and groundwater. Latitude and longitude coordinates of wells and surface water intakes are required, but the location for groundwater withdrawal may be the centroid of a well field, rather than the location of each individual well. Data are checked against the previous year's water use amount and against the permitted amount.

Massachusetts has the legal authority to permit or register water withdrawals. The state requires water use reporting for public, industrial, commercial, irrigation, and livestock water use, but not for domestic and power water use. Monthly water use is reported annually for all users exceeding 100,000 gallons per day, except for public use, where use is reported if at least 25 people or 15 connections are served. Laws are applicable statewide and are the same for surface and groundwater. Latitude and longitude data are recorded for wells and surface-water intakes for public supply, commercial, industrial, irrigation, and livestock water users. Water withdrawal data are checked by the state against the previous year's withdrawal amount, and background data, such as the presence of new wells, new owners, and corrections to the latitude and longitude, are also checked.

Michigan has the legal authority to register water withdrawals. Water use data are reported to the state for public water supply, industrial, and power use, but not for commercial, domestic, irrigation, and livestock with use. Irrigation is currently a split category; golf course irrigators are required to report annual water use, while agricultural irrigators are not. Monthly data are reported for any facility capable of withdrawing 100,000 gallons per day in any 30-day period. Laws are applicable statewide and are the same for surface and groundwater. Latitude and longitude coordinates of wells and surface water intakes are not recorded.

Minnesota has a water use permit program and collects monthly water use data through an annual reporting process for all water uses exceeding 10,000 gallons per day or 1 million gallons per year. Laws are applicable statewide and are the same for surface and groundwater. Location data for wells and surface water intakes are stored in the form of township, range, section, and quarter section down to the nearest 10 acres. A state database of reported water withdrawals is updated annually if staff are available. Some data are field checked and a web page (http://www.dnr.state.mn.us/waters/programs/ water_mgt_section/ appropriations/wateruse.html) shows trends in use over time for several water use categories.

Mississippi has a water use permit program for all users of surface water and groundwater. Permits are not required for groundwater wells less than six inches in diameter. Laws are applicable statewide. Annual water use is reported yearly on a voluntary basis. Public water supply withdrawal points are recorded by latitude and longitude, using global positioning system technology, and also are recorded by township, range, and section. All other permitted withdrawal points are indexed by township, range, and section. Withdrawal data are checked by comparison with the corresponding data for the previous year.

Missouri has the legal authority to require users of 100,000 gallons per day or more to register with the state and report usage annually. However, this is largely a voluntary program. The state maintains a database of annual water use by these users. Laws are applicable statewide and are the same for surface and groundwater. Location coordinates are not required, but many users file latitude and longitude coordinates or township, range, and section location. Withdrawal data are checked by comparison with the corresponding data for the previous year and by looking for outlier data values.

Montana has a water permit program for new water uses. This is a one-time permit for all future years and is not annually renewed. The state does not maintain a water use database. Some new water users are required to submit usage data "at the request of the department." There is no requirement to report latitude and longitude of the water withdrawal. Laws are applicable statewide and are the same for surface and groundwater. Reported data, where available, are checked against permit limits.

Nebraska has the legal authority to permit or register withdrawals for surface water use, without the use of a trigger level. Records of locations of surface water points of diversion and groundwater wells are required. Surface water withdrawals in some basins, such as the Republican River basin, are reported to the state. Some information on public water suppliers is collected by the Nebraska Health Department. The state maintains an annual water use database. Laws are different for surface water (prior appropriation) and groundwater (correlative rights). Locations of points of diversion and groundwater wells are required, but they are not stored as latitude-longitude coordinates. An effort is being mounted to obtain the latitude and longitude of public water supply points. The state tracks changes in water withdrawals through time.

Nevada has a water use permit system for all water use categories except domestic water use. The state engineer determines who must report water use, based on the withdrawal amount. Monthly, quarterly, or annual data may be required, depending on the permit. Laws are applicable statewide and are the same for surface and groundwater. The state does not maintain a water use database. Latitude and longitude coordinates of the water withdrawal locations are not recorded.

New Hampshire has the legal authority to register water withdrawal for all water users exceeding 20,000 gallons per day averaged over a seven-day period. Permits are required only for groundwater withdrawals exceeding 57,600 gallons per day. The state collects monthly water use data on an annual basis for all registered and permitted users. The latitude and longitude of water use sites are stored in a specially designed New England Water Use Database System, which besides water use includes points of water discharge, locations of treatment plants and major distribution system facilities, and their linkage with one another so that the movement of water can be traced from point of withdrawal from the natural water system, through the infrastructure water system, to the point of discharge to the natural water system again (see Chapter 7 of this report for more information on this data system).

New Jersey has a water permit system for all water uses and collects monthly water use data for users exceeding 100,000 gallons per day (or capable of pumping 70 gallons per minute). The USGS maintains a water use database for the state that is being updated continually as new data come in. Latitude and longitude coordinates of all withdrawal points are recorded. Water use data are checked for users renewing or modifying a permit and for users in the proximity of a new water allocation permit.

New Mexico has a water permit system for all water uses. All permit holders regardless of level of use are required to report water use to the state on a quarterly or annual basis, except for domestic wells using less than 3 acre-feet per year and most irrigation water rights. All irrigation wells in the Roswell Artesian Basin are metered by court decree, and the owners are required to report usage. The state engineer may assume jurisdiction over water appropriation and use in other areas by "declaring" any groundwater basin with reasonably defined boundaries. Wells in declared basins may only be drilled with a permit, and they can only be drilled by well drillers licensed by the state engineer. Approximately 40 percent of water right holders are noncompliant about reporting their usage. Some reported withdrawals are monitored to ensure they do not exceed water right allocations. The state is entering available site-specific water use data into its database. This database includes the legal description (township, range, section) of the diversion points. The state also maintains separate databases for compiling five-year summaries of water use by category, county, and river basin. Data for the five-year inventories are compared to previously reported values in order to detect reporting errors or significant changes related to economic and population trends. Water use data are analyzed to support regional water demand projections.

New York has three different water use data collection systems depending on location within the state. Public water supply data are collected statewide for all EPA-regulated systems (more than 25 people served or 15 connections). In Long Island, a set of four counties (Kings, Queens, Nassau, Suffolk) have a special data collection program in which all users of groundwater at pumping rates of greater than 45 gallons per minute are inventoried by the state. In the Great Lakes Basin (42 percent of the state), all water users withdrawing more than 100,000 gallons per day, or having a consumptive use of 2 million gallons over a 30-day period are inventoried by the state. The public water supply database stores annual water use data, is updated every three years, and includes latitude and longitude of the water withdrawal points, although some data are missing. The Long Island database contains annual water use data, is updated annually, and is checked by comparison with the previous year's water use. Latitude and longitude are stored for groundwater wells, although the value reported may be the centroid for a well field. The Great Lakes region database stores annual water use data, is updated every two years, and does not contain latitude and longitude of the water withdrawal points.

North Carolina has the legal authority to register water use, and water use reporting is mandatory within a critical capacity area for all users exceeding 100,000 gallons per day. Outside this area, water use reporting is voluntary and is requested of users exceeding 1 million gallons per day. Data on public water supply are collected every five years, and data are collected for other categories of water use through the registration program. Laws are applied the same to surface water and groundwater.

North Dakota has a water use permit program for all uses except those for which the amount used is less than 12.5 acre-feet per year and the use is for domestic, livestock, fish, wildlife, or recreation. For all permitted users, annual withdrawals are reported on individual response forms. The same laws apply to both surface and groundwater and are applicable statewide. The state maintains a water use database, which is updated every year when the response forms are returned to the state. Latitude and longitude coordinates of the withdrawal points are not recorded, but the township, range, and section values are required. Spot checks of the data are carried out by field-checking pumpage rates or electric consumption of the user. Reports showing unreasonable amounts of use for a given category or amounts exceeding the permitted amount are checked.

Ohio has the legal authority to permit or register water withdrawals and collects water use data annually for any facility capable of withdrawing 100,000 gallons per day. The same laws apply to both surface and groundwater and are applicable statewide. Latitude and longitude coordinates of water withdrawal points are recorded. Withdrawal data are checked by comparison with the previous year's values.

Oklahoma has a water use permit program for all categories except domestic water use. Permit requirements apply statewide for both surface water and groundwater, although ground water is considered a property right and surface water is considered to be publicly owned. The Oklahoma Water Resources Board (OWRB) maintains latitude and longitude locations of withdrawal points, which are determined from legal descriptions using conversion programs. Water use data are requested annually for any magnitude of use by all permitted water users; about 60 percent of the annual surveys are returned to the state. The OWRB maintains a water use database that is updated annually. Reported data are not checked against other information; however, surface water withdrawals are subject to annual review. The exception to the OWRB permitting and reporting requirement is for surface water withdrawals in the Grand River basin, which are under the jurisdiction of the Grand River Dam Authority. The GRDA maintains some records of surface water sales in the Grand River basin.

Oregon has a water use permit program and requires water use data of all permit holders. Various trigger levels for reporting are used, such as 15,000 gallons per day for domestic water use of less than 15,000 gallons per day, and 5,000 gallons per day for commercial and industrial water use. Some categories of water use are exempted, such as livestock watering and fisheries management. The reporting requirements are consistent statewide and require monthly water use data to be reported annually. The state locates diversions using township, range, section, quarter section, and quarter-quarter section values, and it is locating significant water withdrawal points using global positioning system (GPS) technology. The USGS and the state are using GPS to record well locations for specific groundwater projects. The water use database is updated continually as new reports are received.

Pennsylvania regulates public water supply only at the state level, with other categories of water use being regulated by the Delaware and Susquehanna River Basin Commissions, within the boundaries of those river basins. The two river basin commissions have the authority to register all water users exceeding 10,000 gallons per day and to permit water uses for all water users exceeding 100,000 gallons per day. The Susquehanna Commission has the authority to regulate all consumptive users of greater than 20,000 gallons per day. The Delaware River Basin Commission has authority to regulate water use for groundwater users exceeding 10,000 gallons per day in a special groundwater use area. The frequency of reporting water use data varies, with the Delaware Commission requiring annual reports and the Susquehanna Commission requiring monthly, quarterly, or annual reports, depending on the user. Some data checking is performed against permit levels and the past year's data as new water use data are stored.

Puerto Rico has the legal authority to permit or register water withdrawals; it maintains a water use database for all water uses regardless of the amount of use. The frequency of reporting water use data varies, depending on the amount of water withdrawn, the use, and the source of water. There is a fee assigned to the user depending on the water use, but the fee does not apply to water used in agricultural activities. Laws are applicable statewide and are the same for surface water and groundwater. Latitude and longitude coordinates of the withdrawal points are not recorded. Data reports are completely revised every three to five years at the time of permit renewal.

Rhode Island has the legal authority to permit or register water withdrawals for public water use only; it does not have this authority for any other usage category. The state requires collection of public water use data for users exceeding 100,000 gallons per day, regardless of whether the source is surface water or groundwater. The state does not maintain a water use database on an ongoing basis.

South Carolina has the legal authority to permit or register water withdrawals for all categories of water use. Data are recorded quarterly, or monthly during times of extremely low stream flow, for users that exceed 3 million gallons in any month (or approximately 100,000 gallons per day). Laws are applicable statewide and are the same for surface water and groundwater. The state maintains a water database on an ongoing basis. Latitude and longitude coordinates of withdrawal points are not required. Data are checked against the previous year's use and are summarized in a data report every one to three years.

South Dakota has a water use permit program for all categories of water use except domestic use. The state collects annual water use data from the largest public water supply systems and from all irrigation users, and every five years the state requests data voluntarily from other water users using questionnaires. Laws are applicable statewide and are the same for surface water and groundwater. Well and surface water intake locations have township-range-section values as part of the permit application, and these locations are converted to latitude and longitude by the state or USGS. Irrigation water use data are used collectively to determine whether water is available from a particular source for further appropriations, and individual irrigation water use data are used to determine whether an existing right is still active or subject to cancellation due to nonuse. No metering is required. The state maintains a database, updated annually, of irrigation withdrawals. The USGS maintains a site-specific database of public water supply, industrial, thermoelectric, and irrigation data. Sometimes, during enforcement actions, water use data are checked against power consumption data.

Tennessee does not have the legal authority to permit or register water withdrawals. Public water use data are reported to the state, and the USGS maintains a public water supply database. Public water supply data are recorded monthly and are reported annually for all surface water withdrawals and for groundwater withdrawals for systems serving more than 50 people. Latitude and longitude coordinates of the withdrawal points are recorded both for wells and surface water intakes.

Texas has legal authority to permit surface water use throughout the state, but groundwater use permits are required only in particular groundwater conservation districts. Water use data for municipal and industrial uses have been voluntarily submitted to the state for many years, but beginning in November 2001, water use data collection is mandatory for both surface water and groundwater users. Monthly water withdrawals are requested, but often only annual data are supplied. There has been no specified trigger level for water use data collection since data collection has been voluntary. The state maintains a water use database, which is updated annually. The locations of groundwater wells and surface water intakes are shown on maps. As part of a statewide water availability study, latitude and longitude coordinates of all permitted surface water diversion points are being determined. Changes in water use from year to year are tracked in a quality assurance process, and revisions to the data are made when necessary.

Utah has the legal authority to permit or register water withdrawals for all categories of water use. Annual water use data are collected for all water users exceeding 20 acre-feet per year, except for domestic, livestock, and irrigation use. Irrigation water use is estimated by the USGS using pumpage inventories and electric consumption records. These data are provided to the state and are published annually. A special emphasis in data collection is made for areas where groundwater management plans have been developed. Surface water withdrawals are monitored by river basin commissioners or local water entities. Well locations and surface water intakes are located by township-range-section but not by latitude and longitude. Water use data are checked and updated as water rights change and as large changes in withdrawals are noted from the previous year. The data are also checked and updated using field reviews.

Vermont does not have the legal authority to permit or register water withdrawals. Water use data are not reported to the state. Water use data for Vermont have been compiled by the USGS and incorporated within the New England Water-Use Data System in the same format as that for New Hampshire (see Chapter 7 of this report for more information on this data system).

Virginia maintains a water use register for all categories of water use except domestic. Monthly data are reported annually to the state for users whose average withdrawal rate exceeds 10,000 gallons per day for any single month and for irrigators whose use exceeds 1 million gallons per month. The procedure is applicable statewide and it is the same for surface water and groundwater, with the exception of two coastal Ground Water Management Areas, where withdrawals of greater than 300,000 gallons per month must be reported. The locations of wells and surface water intakes must be shown on a map, and the latitude and longitude are requested if known by the user.

Washington has the legal authority to register the construction of wells, and permits are required for any use other than single-family domestic withdrawals. Water use data are not reported to the state.

West Virginia does not have the legal authority to permit or register water withdrawals. Public water use data are not reported to the state.

Wisconsin has the legal authority to permit or register water withdrawals for public water supply, industrial, and power use only; it does not have this authority for commercial, domestic, irrigation, or livestock use. Water use data are reported to the state annually for all public supplies, for industrial water users exceeding 100,000 gallons per day, and for all thermoelectric power facilities. The laws are applicable statewide and are the same for surface water and groundwater. A water use database is maintained on an ongoing basis, but whether latitude and longitude of water withdrawal points are stored in the database is unknown. Water use data for public water supplies are checked against the previous year's water use.

Wyoming has a prior appropriations doctrine that requires permits for beneficial use of water. The Wyoming State Engineer's Office maintains a database of permit information, which is useful for developing USGS water use estimates. Permits are required both for surface and groundwater use, although a usage rate of less than 25 gallons per minute is considered domestic water use and has a simpler permit process. The Wyoming Water Development Commission conducts a biannual survey of public water systems (see http://wwdc.state.wy.us/watsys/2000/raterept.html for the 2000 report), and in alternate years it conducts a similar survey of agricultural water use. These surveys contain annual water use data and a significant amount of ancillary information about each water user. Latitude and longitude data have been collected for public water supplies but generally not for other water users.

Appendix B

Biographical Sketches of Committee Members

DAVID R. MAIDMENT, *Chair*, is the Ashley H. Priddy Centennial Professor of Engineering and director of the Center for Research in Water Resources at the University of Texas at Austin. He is an acknowledged leader in the application of geographic information systems (GIS) to hydrologic modeling. His current research involves the application of GIS to floodplain mapping, water-quality modeling, water resources assessment, hydrologic simulation, surface water–groundwater interaction, and global hydrology. He is the coauthor of *Applied Hydrology* (McGraw-Hill, 1988) and the editor-in-chief of *Handbook of Hydrology* (McGraw-Hill, 1993). From 1992 to 1995 he was Editor of the *Journal of Hydrology*, and he is currently an associate editor of that journal and of the *Journal of Hydrologic Engineering*. He received his B.S. degree in Agricultural Engineering from the University of Canterbury, Christchurch, New Zealand, and his M.S. and Ph.D. degrees in civil engineering from the University of Illinois at Urbana-Champaign.

A. ALLEN BRADLEY is an associate professor of civil and environmental engineering at The University of Iowa and a research engineer at IIHR Hydroscience & Engineering. His research interests are in the areas of hydrology and hydrometeorology, including flood and drought hydrology, hydroclimate forecasting, and water resource applications of remote sensing. He received his B.S. in civil engineering from Virginia Tech, an M.S. in civil engineering from Stanford University, and a Ph.D. in civil and environmental engineering from the University of Wisconsin.

MICHAEL E. CAMPANA is director of the Water Resources Program and professor of earth and planetary sciences at the University of New Mexico. His current interests are hydrologic system–aquatic ecosystem interactions, regional hydrogeology, environmental isotope hydrology, and the hydrology of arid and tropical regions. He teaches courses in water resources management, hydrogeology, subsurface fate and transport processes, environmental mechanics, and geological fluid mechanics. He was a Fulbright Scholar to Belize in 1996. Dr. Campana received a B.S. in 1970 in geology from the College of William and Mary, an M.S. in hydrology in 1973, and a Ph.D. in hydrology in 1975 from the University of Arizona.

BENEDYKT DZIEGIELEWSKI is an associate professor of geography at Southern Illinois University at Carbondale and executive director of the International Water Resources Association. His two main research areas are water demand management (urban water conservation planning and evaluation, water demand forecasting, modeling of water use in urban sectors) and urban drought (drought planning and management; measurement of economic, social, and environmental drought impacts). He is editor-in-chief of *Water International* and is an honorary lifetime member of the Water Conservation Committee of the American Water Works Association. He received his B.S. and M.S. in environmental engineering from Wroclaw Polytechnic University, Wroclaw, Poland, and his Ph.D. in geography and environmental engineering from Southern Illinois University.

N. LEROY POFF is an assistant professor in the Biology Department of Colorado State University. Dr. Poff received a B.A. in biology from Hendrix College in 1978, an M.S. in environmental sciences from Indiana University in Bloomington, Indiana, in 1983, and a Ph.D. in stream ecology from Colorado State University in 1989. His primary research interests are in stream and aquatic ecology. Dr. Poff currently teaches an introductory course in biology and two advanced courses in aquatic ecology.

KAREN L. PRESTEGAARD is an associate professor of geology at the University of Maryland. Her research interests include sediment transport and depositional processes in mountain gravel-bed streams; mechanisms of streamflow generation and their variations with watershed scale, geology, and land use; hydrologic behavior of frozen ground; hydrologic consequences of climate change; and hydrology of coastal and riparian wetlands. She was a member of the NRC/CGER/BRWM Committee for Yucca Mountain Peer Review: Surface Characteristics, Preclosure Hydrology, and Erosion. She received her B.A. in geology from the University of Wisconsin-Madison, and her M.S. and Ph.D. in geology from the University of California, Berkeley.

STUART S. SCHWARTZ is associate director of the Water Resources Research Institute (WRRI) of the University of North Carolina. His research interests include watershed management, risk-based reservoir operation, and the use of probabilistic forecast information in the planning and operation of water resource systems. Before joining WRRI, he directed the Section for Cooperative Water Supply Operations on the Potomac (CO-OP) at the Interstate Commission on the Potomac River Basin, and he was an associate hydrologic engineer at the Hydrologic Research Center in San Diego, California. He received B.S. and M.S. degrees in biology-geology from the University of Rochester and a Ph.D. in water resource systems analysis at the Johns Hopkins University.

DONALD I. SIEGEL is a professor of geology at Syracuse University, where he teaches graduate courses in hydrogeology and aqueous geochemistry. He holds B.S. and M.S. degrees in geology from the University of Rhode Island and Pennsylvania State University, respectively, and a Ph.D. in hydrogeology from the University of Minnesota. His research interests are in solute transport at both local and regional scales, wetland-groundwater interaction, and paleohydrogeology. He was a member of two NRC committees: Committee on Techniques for Assessing Ground Water Vulnerability and Committee on Wetlands Characterization.

VERNON L. SNOEYINK is the Ivan Racheff Professor of Environmental Engineering at the University of Illinois. His primary areas of research are the physical and chemical processes for drinking water purification, in particular the removal of organic contaminants by activated carbon adsorption. In 1980, he coauthored the textbook *Water Chemistry* (Wiley and Sons). He has been a trustee of the American Water Works Association Research Foundation and president of the Association of Environmental Engineering Professors. He is now a member of the editorial advisory board of the *Journal of the American Water Works Association* and vice-chair of the Drinking Water Committee of the U.S. Environmental Protection Agency's Science Advisory Board. He was elected to the National Academy of Engineering in 1998. He has been a member of several NRC committees and chaired the Committee on Small Water Supply Systems. He received his B.S. and M.S. degrees in civil engineering and his Ph.D. in water resources engineering from the University of Michigan.

MARY W. STOERTZ is an associate professor of hydrogeology at Ohio University, Department of Geological Sciences. Her area of specialty is stream restoration, especially restoration of channelized rivers and streams polluted by acid mine drainage. She founded the Appalachian Watershed Research Group at Ohio University, which has the mission of restoring desired functions of watersheds subject to mining, sedimentation, and flooding. She directs the multidisciplinary research arms of the Monday Creek Restoration Project and the

Raccoon Creek Improvement Committee. Dr. Stoertz received her B.S. in geology from the University of Washington and her M.S. and Ph.D. in hydrogeology (with a minor in civil and environmental engineering) from the University of Wisconsin-Madison.

KAY D. THOMPSON is an assistant professor at Washington University, Department of Civil Engineering. In her research, she investigates properties of subsurface materials for groundwater studies, develops methods for subsurface characterization, assesses the risks of hydrologic dam failure, and consults on minimizing environmental impacts during development. Dr. Thompson received a B.S. in civil engineering and operations research in 1987 from Princeton University, an M.S. in 1990 from Cornell University, and a Ph.D. in 1994 in civil and environmental engineering from the Massachusetts Institute of Technology.